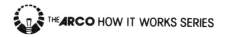
THE**ARCO** HOW IT WORKS SERIES

AND THE CREATION OF NEW LIFE

Christopher Lampton

ARCO PUBLISHING, INC.
NEW YORK

For Lynn and Karen—
who wanted copies of their own.

Published by Arco Publishing, Inc.
215 Park Avenue South, New York, N.Y. 10003

Library of Congress Cataloging in Publication Data

Lampton, Christopher.
   DNA and the creation of new life.

   (The Arco how-it-works series)
   Includes index.
   1. Genetic engineering. 2. Recombinant DNA.
I. Title. II. Title: D.N.A. and the creation of new
life. III. Series.
QH442.L35   575.1   82-6874
ISBN 0-668-05396-8 (Cloth edition)   AACR2
ISBN 0-668-05401-8 (Paper edition)

Printed in the United States of America

10   9   8   7   6   5   4   3   2   1

Contents

Introduction: The Genetic Revolution   *v*

PART ONE
The Quest for the Gene   1

PART TWO
Shaping the Gene   29

PART THREE
Selling Genes   77

PART FOUR
The Engineers of Life   117

Index   133

Introduction:
The Genetic Revolution

Near the turn of the seventeenth century, when Galileo trained his telescope on the skies, he saw things that no human being had seen before—and our view of the universe was irrevocably changed.

When Max Planck announced in 1900 that energy came in tiny units he called *quanta*, he neatly demolished everything physicists had believed about the ultimate nature of reality—and in so doing Planck gave them a view of reality more breathtaking by far than any they had imagined.

Science moves by fits and starts, but when it gains momentum it sometimes is hard to stop; it is as likely to vault whole canyons of ignorance as to plug tiny holes of doubt. When a scientific revolution is in full ferment, discoveries come hard and fast; sometimes they contradict everything that has gone before. Suddenly, we are forced to see the universe through new eyes and, in so doing, we may see more than we have ever seen.

Physics, in modern times, has been blessed with two such revolutions. One, spearheaded (if not begun) in the second half of the seventeenth century by Isaac Newton, showed us the very tick and whir of the clockwork universe; the other, early in this century, showed us that this was no ordinary clock.

Biology has been less prone to such violent upheavals, though ripples of change have from time to time passed through its fabric. Perhaps because the biologist has traditionally dealt in less fundamental stuff than the physicist, these ripples have been more localized, less likely to alter the fabric as a whole.

Until now.

In the wake of the revolution in physics has come a related revolution in biology. Biologists, like physicists, now deal in the particles of which matter is made; by understanding these particles

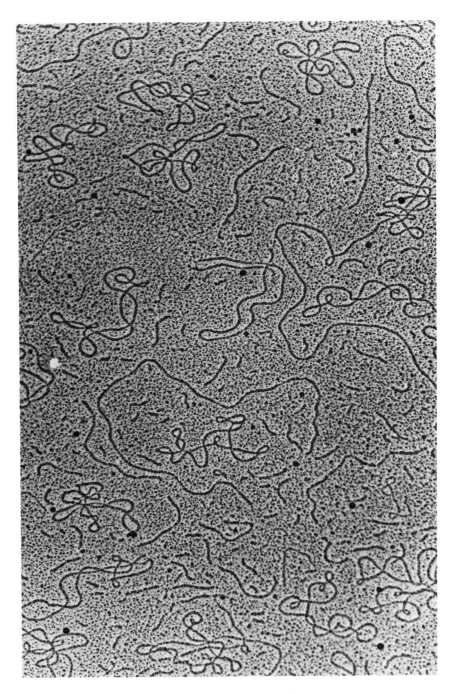

Here is what everyone was looking for—DNA. This is an electron microphotograph of DNA. *(courtesy Genex Corporation)*

an understanding of life itself has evolved—one that is leading us into a new age.

This is no trivial thing. Life is mysterious, miraculous, but it is understandable on the same terms as everything else. This is the lesson of molecular biology, the relatively new discipline that has grown out of the fusion of physics and the life sciences. Life is not only understandable, but also, if we can find the right tools, we can pierce to the very heart of the living organism and change it as we will.

And now we have the right tools.

We have entered the age of genetic engineering, the age of biological manipulation. We have ceased to be slaves to the mindless processes that have driven every living organism since life first appeared in the primeval sea and have become instead the masters.

It is now within our power to alter the chromosomes of bacteria in any of a number of useful ways. We can engineer microorganisms that will turn out nearly unlimited quantities of chemicals not usually found in microorganisms—human hormones, for instance. We can design artificial molecules just like those once thought to be the exclusive province of living creatures, and we can design other molecules that have never existed before in nature. Soon we will be able to perform similar manipulations on the chromosomes of plants, creating self-fertilizing crops, or crops that will not require insecticides. Within a few decades we may even be able to engineer the chromosomes of animals in much the same way.

Within a century, it is possible that we will have the knowledge and the ability to cure genetic diseases in human beings; that we will be able to design the genetic inheritances of our children before they are born; that we will be able to redesign the very nature of the human organism.

No one can say if all these things are possible, but already we can speak, in general terms, about how they will be done, if they can be done at all.

Thirty years ago, the way that genetic information was carried from parent to child was not yet known. Today, we not only know how that information is written, but we also know how to rewrite it. In those thirty years, molecular biology has progressed over a staggering distance—and yet, with the tools available to biologists today, we should cover as great a distance in the next five to ten years as in the past thirty. Our knowledge is expanding along a geometric curve; anyone who doesn't keep up with what is being learned today may find himself or herself swamped by the sheer weight of information that we will accrue tomorrow.

This book follows the road taken by biologists over the past century in their attempt to understand the molecular nature of life: specifically, the nature of the factor of inheritance called the gene. Part I talks about the search for the gene and the unraveling of the genetic code. Part II explores the technology—the so-called recombinant-DNA process—that allows the gene to be manipulated, and the controversy that has surrounded that technology. Part III examines the industry that has sprung up around genetic engineering, and Part IV takes a look at the future and the possibilities for turning the gene splicer's scalpel on the chromosomes of human beings.

The engineering of life is a technology that holds out great promise, but it also is a technology with profound moral and social implications. It is a technology that may change the way you live in the future, but its implications must be understood if we are to have any future at all.

The engineers of life have arrived. It is up to the rest of us to decide what we want them to do.

PART ONE

The Quest for the Gene

THE CELL

The key to it all is the cell. This sometimes hardy, sometimes fragile glob of protoplasmic jelly is the unit of life, and in a sense it is the only living thing there is. Dismantled, it can contribute its working machinery to what biologists call "cell-free systems," but these systems, though they can mimic the activity of life, are not truly alive. Viruses, which are not precisely cells, do a brilliant job of pretending to be alive, but in order to make their mimicry complete they must play the role of microscopic pirates, hijacking the machinery of full-fledged cells in order to reproduce their kind.

The first cell must have come into existence about three or four billion years ago, no more than a billion years after the earth formed out of molten elements in orbit around the sun. By some theories, the cell may have—ironically—predated the existence of life itself, forming by chemical processes in the primeval seas, so that when the first primitive organic molecules cast about for some form of protective cover, it was only natural that they should take up residence in these ready-made, pseudo-organic bubbles.

Even today, all life on this planet, with the quasi-exception of viruses, is made of cells. In the human body there are trillions of them, none (with the exception of the egg cell of the female) large enough to be seen with the naked eye. Each of us is a colony of cells, though long ago the individual colonists gave up their autonomy in favor of the totalitarian rule of the glandular and nervous systems. Not one of these cells could survive long on its own; as in society in general, specialization is the order of the day. But each cell carries within itself, complete and entire, the fabulously complex instructions for running (and assembling) the organism as a whole.

1

This is the irony on which the engineering of life depends. The secret of life (which is no longer a secret, at least not fully) is contained in every cell of every living organism. There are *quintillions* of copies, spread across the landscape like yesterday's lottery tickets. It is contained a billion times over in every blade of grass. It can be found in the cells of dandelions and Lhasa apsos, elephants and jellyfish. To biologists, who have cracked the code in which the secret is written, each cell is a kind of library, replete with the instructions for creating life itself, instructions we could follow if only we had the right tools.

We may not have the proper tools for creating life, but we do have the tools for altering it, for entering certain bacterial cells and editing the messages contained within, for mixing together genes from different species and creating hybrid organisms—in essence, for creating new forms of life from old, if not from scratch.

We stand on the verge of a new age. Does this seem an extreme statement? You must understand that we have in our hands a technology that will reshape the future, but it is a technology built on an edifice of more than a century's research—and thousands of years of speculation—on the nature of the living organism.

WHAT IS LIFE?

Perhaps more to the point, how does life differ from things that are not alive?

Cast a billiard ball against the sideboard of a velvet-covered billiard table and it will respond in certain predictable ways. If you know the speed at which the ball is traveling, the angle at which it is going to hit the sideboard, and the elasticity of the ball, it is possible—in theory, at least—to predict the precise angle and speed with which the ball will rebound. If it should collide with other balls, we can similarly predict how much momentum will be imparted to each and in what directions they will move.

Billiard balls obligingly obey the laws of physics, which makes their behavior fairly simple to account for. Since ballistic missiles obey the same laws, an understanding of the behavior of billiard balls will generally prepare one for calculating the trajectories of ICBMs. And that, of course, is what physics is all about: defining the ways in which matter and energy interact in such general terms that the same rules can be applied to all systems that might come under investigation. The same laws derived from the study of billiard balls and ballistic missiles can be applied to black holes and quasars, though such astronomical phenonema lie far beyond the reach of our calipers and measuring rods.

At first glance, however, living organisms would seem to present something of a special case. They flagrantly defy the law of entropy, which says that physical systems become more and more disordered with time; in fact, living organisms actually become increasingly ordered with time. And chemical reactions seem to take place within biological systems that have no counterpart in the rest of nature.

The idea that life somehow was separate from nonlife—that it obeyed different physical laws—is called *vitalism*, and it dominated biology until sometime in the past century. Even the great biochemist Louis Pasteur (1822–1895) championed the vitalistic cause, though in the end the vitalists lost out to the mechanists, those who believed that the organic and the inorganic—the living and the nonliving—were all of a piece.

As in fact they are. Living systems do indeed become more highly organized with time, but they do this only at the expense of greater disorganization in the world around them; thus the laws of entropy are obeyed. The chemical reactions that take place in living organisms are different from those in the nonliving world, it is true, but the differences are ones of complexity, not of principle. No physical laws are broken, though several are sorely stretched.

The secret of living organisms lies in the structure of the molecules and atoms that make them up; and the key to that structure is a particular molecule, called the *chromosome*, which contains certain molecular sequences called *genes*.

And these molecular sequences are at the center of the great work of twentieth-century biology.

LIFE BEGETS LIFE

There is a continuity to life that would be difficult for even the most blinkered of observers to overlook. Not only does like tend to beget like—fish beget fish and fowl beget fowl—but also, like tends to beget similar. Children look like their parents, or at least like their grandparents (or an aunt or an uncle or somebody back down the line). Certain diseases, like hemophilia, tend to run in families (especially in those such as the royal families of Europe, where intermarriage has been historically common). It takes only scant knowledge of biology to recognize that more is being handed down from generation to generation than just the family heirlooms. Something invisible but very much alive must be passed down as well—and that something must contain the essential blueprint of the newborn individual who inherits it, a blueprint that says that a given child is destined to develop his father's nose and his mother's chin, his grandfather's blond hair and his great-aunt's blue eyes.

Aristotle (384–322 B.C.), that great propounder of theories and detester of experimentation, believed that it was something in the blood, a kind of liquid inheritance that blended together to make each child a composite of everyone who had gone before, a little bit Mom and a little bit Dad and a little bit everyone else. Aristotle was on the right track, but as usual he harbored some crucial misconceptions.

Chief among these was the idea that inheritance represented a kind of blending of ancestral traits, the way a milk shake represents the blending of vanilla ice cream and chocolate syrup. This was a misconception that persisted into the present century—and to some extent has yet to die out—but fails to do justice to the subtlety of the processes involved.

Gregor Mendel was perhaps the first person to grasp the true nature of biological inheritance. Mendel, a monk, raised generation after generation of pea plants in the garden of his monastery at Brunn (now part of Austria) in the nineteenth century, meticulously observing the way in which certain characteristics of the plants were passed from parent to child.

Mendel's findings were striking, almost revolutionary in their simplicity. No blending of characteristics took place: The offspring of a plant with red blossoms and a plant with white blossoms was not a plant with pink blossoms, but either a plant with red blossoms or a plant with white blossoms. These two alternatives occurred among the offspring according to well-defined mathematical ratios (so well defined, in fact, that some modern researchers have suggested that Mendel may have fudged his statistics somewhat).

It was as though some kind of unit of heredity were passed on from parent to offspring, each parent contributing a single unit for each hereditary characteristic. If the offspring received two differing units for a single characteristic (a unit for red blossoms from the mother, say, and a unit for white blossoms from the father), only one could express itself—that is, only one could have an effect on the physical appearance of the organism. The one that expressed itself Mendel termed the *dominant* characteristic, the other the *recessive* characteristic. Either of these characteristics then could be passed on to the third generation; a recessive characteristic, however, could be expressed only if it were inherited in tandem with an identical recessive characteristic. For instance, a plant that received two units for red blossoms (one from each parent) would itself have red blossoms; so would a plant that received a unit for red blossoms and a unit for white blossoms. Only a plant that received two units for white blossoms would express that characteristic, because white blossoms are recessive.

What were these mysterious units that carried with them the blueprints for generations yet unborn? Mendel had no idea; his evidence for the existence of such units was compelling but entirely indirect. For nearly a half-century after Mendel published the results of his experiments in a small scientific journal, his findings went unrecognized, and no one attempted to pursue this line of research. Then, in the year 1900, three biologists independently rediscovered Mendel's work and brought it to the attention of the world. Mendel's "units" soon were rechristened *genes*, an appellation derived from the term *genetics*, the name earlier given to the study of biological inheritance.

What are these genes? This is not an easy question. The answer was to come from several independent lines of research, one of them the so-called classical genetics of which Mendel was the almost inadvertent founder. At the forefront of this field in the early twentieth century was a scientist named Thomas Hunt Morgan, who studied the genetics of fruit flies in much the same way that Mendel had studied pea plants. Morgan, however, made a discovery that Mendel seemed to have overlooked.

In his studies of peas Mendel showed that inherited characteristics were passed on from generation to generation in an essentially independent assortment—that is, the gene that coded for color of blossom, say, was inherited independently of the genes for seed texture or height of stalk. Morgan, on the other hand, observed that certain characteristics in the fruit fly were mysteriously linked—that is, they were often passed on in conjunction with one another, more often than the laws of chance would allow. Morgan conjectured that there must exist a physical attachment between these particular genes, that they were chained together like beads on a string.

When Morgan enumerated the characteristics of the fruit fly, he found that they fell into four linked groups, three large and one small. Genetic characteristics within each group were inherited together more often than not—some considerably more often. Characteristics in different groups were inherited together by random chance only; there was no apparent linkage between them. If one imagined that fruit-fly genes somehow were chained together, then they would seem to form four chains, three large ones and one small one. Was there physical evidence that such chains existed?

As it happens, there was. Research in cell biology had earlier shown that there are structures in the center of the living cell known as chromosomes (so named because of the way they absorb the colored dyestuffs that scientists use to make them visible). Locked inside a small compartment within the cell called the nucleus, the chromosomes are threadlike bodies almost invariably grouped in pairs, the number of pairs varying from organism to organism. Human beings have twenty-three such pairs. Fruit flies have four—one small and three large. Was this coincidence?

Almost certainly not. The evidence for the chromosomes being the site of the genes was overwhelming. They were ubiquitous— they existed in every cell of every organism examined—and they seemed of prime importance to the cell. When parent cells split in two to form progeny cells, they carefully duplicate their chromosomes and pass on identical copies to both of the new cells, as though they contain something quite precious.

But of what were the chromosomes made? How might they carry the blueprint of life from generation to generation?

Those, alas, were not questions for geneticists. And so, at some time in the first half of this century, the quest for the gene passed from those who analyzed the statistics of inheritance to those who studied the atoms and molecules of which the living cell is made: the molecular biologists.

THE SMALLEST REALM

With the passing of vitalism, it became obvious that biology is intimately tied to physics, that the former is in fact no more than a special case of the latter—the physics of living organisms, as it were. Just as physics in the past century has pressed ever inward toward the ultimate nature of matter, so biology has moved more and more into the realm of the very small, the realm of atoms and molecules.

All matter is made up of *atoms*. Is anyone so innocent of science that he or she finds this startling? The atom, as we have come to know it in the past eighty years or so, is an elaborate conglomeration of so-called subatomic particles, mysterious pinpoints of substance which may in turn be made up of even smaller particles— subsubatomic particles, if you will. For our purposes in this book, however, it is enough to imagine atoms as featureless balls, gray spheroids of varying size. The size of the atom, of course, depends on what kind of atom it is. Atoms of the element hydrogen are extremely small, while atoms of uranium are relatively large. (It should be added that all atoms are, by human standards, nearly infinitesimal, too small to be seen with conventional optical micro- scopes.)

Atoms are rarely found in isolation. Instead, they link together into structures called *molecules*, atomic chains that range from two to several million atoms in length. These molecules are held to- gether by what physicists call *bonds*, natural connections that form between one atom and the next. The nature of these bonds is complex, but they can be envisioned as a kind of sticky spot on the surface of the atom, to which other atoms can cling. These bonds come in various strengths. Certain bonds, especially those called hydrogen bonds, are weak, easily broken, rather like velcro or rubber cement. Other bonds are more like epoxy, broken only with considerable difficulty.

How many bonds an atom can make depends on the nature of that atom; every type of atom has its characteristic number. At any given moment, most of these bonds will be filled, though they can be broken and the molecular structures of which they are part rear- ranged as often as nature and circumstance deem necessary. Some atoms, such as hydrogen, can make only one bond. Others can make several—carbon will bond with as many as four other atoms simul- taneously. Still others are reluctant to bond at all. The formation of these bonds requires energy; when the bonds are broken, that energy is released.

In a convenient scientific shorthand, we can refer to the arrang- ing and rearranging of molecules as *chemical reactions*. When sepa- rate atoms come together to form molecules, or when molecules are broken apart into separate atoms, or when various molecules rearrange themselves to form new molecules, we say that a chemical reaction has taken place. Such reactions can be spectacular—clear liquids can be made to take on vivid colors; seemingly inert substances can be induced to fizz violently (as when an antacid tablet is dropped in water) or even explode.

Life is a series of such chemical reactions, most of them taking place within the cell. Molecules of food are chemically dismantled and rearranged in various useful ways, just as an old building might be torn down and its bricks used to construct more buildings. Some of these molecules are used to build new cell structures, or to replace old structures that have fallen into disrepair. Still other molecules are cracked open to release their energy contents, much as an automobile engine breaks down the hydrocarbon molecules in gasoline to release the energy that propels the car. Once the cellular energy has been released, it can be stored in other, newly constructed molecules, for release at a later time—or it can be put to use driving the machinery of the cell.

When these chemical reactions within the cell cease, life ceases. An inert cell is a dead cell.

Chemical reactions, if they are to take place spontaneously, require heat (which is, after all, nothing more than the random motion of molecules). A common chemical activity within the cell is the construction of large molecules, called *polymers*, from smaller molecules, called *monomers*. For this process to take place, the smaller molecules have to come together at precise angles; a particular series of bondable atoms on one molecule must come into contact with a particular series of bondable atoms on the other in such a way that an attachment can be made. Otherwise, the two molecules will simply bounce apart and the larger molecule will not form. If enough random collisions take place between molecules, eventually the right combination occurs and the reaction proceeds. The tiny monomers link together into a massive polymer.

At body temperature—the temperature at which, obviously, reactions must take place within the cell—there isn't enough molecular motion for these reactions to occur, at least not in the quantities required. Even if the temperature should somehow be increased, the effect on the delicately structured life processes of the cell would be disastrous—reactions would proceed out of control; the chemical environment of the cell would become chaos.

Fortunately, chemical reactions don't have to take place spontaneously; they can be brought about deliberately and specifically. And the key to this process is a type of molecule called the *enzyme*.

ENZYMES IN ACTION

The existence of enzymes has been recognized for centuries, though only in the past hundred years has their nature been understood. The purpose of enzymes is simple: They promote chemical reactions. And they function best at body temperature.

Enzymes are polymers—long molecules strung together out of smaller molecules; specifically, enzymes belong to the class of polymers known as *proteins*. The smaller molecules they are constructed from are called *amino acids*.

In the human body there are twenty different amino acids, each similar in form yet with its own unique structure. These amino acids link together like molecular Tinker Toys to form an almost infinite variety of larger molecules, some of them containing hundreds of thousands of atoms, each molecule differing from other such molecules in the specific order in which its amino acids occur.

Every organism contains hundreds, even thousands of different kinds of enzymes, each built up from a distinct sequence of amino acids, each designed to promote a specific chemical reaction. They participate in these reactions without themselves being changed by them, a process known as *catalysis*. (Thus, enzymes sometimes are referred to as catalysts.)

Imagine an enzyme as a kind of molecular machine, in the sense that a lever or a wheel or a wedge is a machine. Rather than being a simple string of atoms, with amino acid following amino acid like the beads in a necklace, each enzyme has a distinct, three-dimensional shape, the chain of amino acids folded back on itself in an elaborate piece of molecular architecture. A particular enzyme might look (if it could be seen with any clarity) like a mass of coral or a chunk of frozen tapioca or even a submicroscopic corkscrew.

How does an enzyme promote chemical reactions? Simply put, it has the right shape—it provides a custom-made surface on which the reaction can take place. An enzyme is a kind of molecular pegboard; pockmarking its surface is a series of indentations called *active sites*. Just as the holes in a pegboard are designed to fit specific kinds of pegs—a round hole for round pegs, a square hole for square pegs, a triangular hole for triangular pegs—so are the active sites on an enzyme designed to fit specific molecules. Molecules, of course, have shapes far more sophisticated and varied than those of ordinary, geometric pegs, so the active sites on the enzymes consequently are more complex.

Consider an enzyme designed to promote a reaction between two specific molecules. Ordinarily, these molecules would be brought together only at random and would not react unless lined up in precisely the right way. At body temperature, this would happen rarely. The active sites on the enzyme, however, are specifically designed to accommodate the shapes of the two reacting molecules, just as the holes in a pegboard are keyed to the shapes of the pegs. Tossed randomly against the enzyme by the gentle molecular motion of the cell, the reacting molecules slide easily into these active sites, where they are captured by weak hydrogen bonds. As they slip into place, they are automatically reoriented into exactly the positions necessary for the reaction to take place. Bondable atom comes into contact with bondable atom. New bonds form, old bonds are broken, the molecules are rearranged. The altered molecules then are released and the active site is freed for further catalysis.

This, of course, is only one possibility. The role of the enzyme may be simply to hold one molecule steady long enough for the other molecule to skewer it; or the enzyme might break one molecule in two, passing the resulting fragments to other enzymes for further reactions.

Life is hectic at the cellular level. In the fashion described above, a single enzyme can catalyze more than a billion chemical reactions per minute. And all of the reactions will be of a specific kind—the kind controlled by that specific enzyme.

How convenient! If the cell is the unit of life, and if chemical reactions are the mechanism by which it functions, then enzymes, by controlling those reactions, ultimately control the entire organism of which the cell is a part. And it follows that whatever controls the enzymes controls everything—cells, organs, living creatures. By their very presence, enzymes imply some kind of guiding intelligence in the cell, a directing force; when we say that enzymes are "designed" to promote specific reactions, the implication is of deliberate purpose. Enzymes are too specific—and, generally, too well timed—to come about by chance. Something must be in charge of enzyme production and therefore of cellular activity in general. But what?

What indeed?

THE MATTER AT HAND

The substance of the gene was discovered in 1869, though at the time no one knew exactly what it was. Swiss biochemist Friedrich Miescher extracted it from the remnants of pus cells found on used bandages appropriated from a hospital. He called it *nuclein*, because it came from the nuclei of the cells. Later it was renamed *nucleic acid*.

There are two kinds of nucleic acid—one containing molecules of ribose sugar and the other containing molecules of deoxyribose sugar. Hence we call the one *ribonucleic acid* and the other *deoxyribonucleic acid*—*RNA* and *DNA* for short. That either RNA or DNA might be material of the gene was considered unlikely until the mid-1940s. The nucleic acids tend to form extremely large molecules—the largest, on average, in the cell—but they are also monotonously repetitive. Both DNA and protein were long ago shown to be present in the chromosomes, but while protein molecules are constructed of some twenty different amino acids, the nucleic acids are made up of only five different smaller molecules— the so-called nucleotides. If the genes carried information from parent to child, how could this information be transcribed on as uncomplex a molecule as DNA? We can draw an analogy here with the alphabet. In ordinary written English, information is carried by twenty-six different letters. Twenty amino acids could probably carry information in similar fashion, but not by five nucleotides (or so it seemed in the first part of this century). Furthermore, early experiments indicated that the five nucleotides occurred in roughly the same amounts within the nucleic acid molecules; there was no pattern or structure to their distribution. (This was later shown not to be precisely the case, though it is not far from the truth.) Without structure, how could information be carried? Protein molecules, on the other hand, were rich with structure. Clearly the substance of the gene was protein; nothing else was even worth considering. The nucleic acids, on the other hand, were assumed to be purely structural, the binding that holds the protein "book" together.

And yet experimental evidence gradually eroded the protein hypothesis. In the early 1940s, Canadian-born biochemist Oswald Avery actually isolated the genes of bacteria in a test tube—and they were not protein at all.

How did he do this?

His research involved the bacterium that causes lobar pneumonia, which comes in two varieties: one with a rough coat on its surface and one with a smooth coat.

The difference is crucial. The bacteria with the smooth coats are virulent, even deadly, while the bacteria with the rough coats have lost, through some evolutionary quirk, the ability to cause disease.

In 1928 biologist Russell Griffith discovered that the harmless, rough-coated bacteria could be rendered virulent by exposure to dead, smooth-coated cells. Some substance, he reasoned, must have passed from the dead cells to the living, a substance that carries the genetic instructions for causing pneumonia. Could that substance be the gene? Almost certainly it was.

Oswald Avery had no particular bone to pick as to whether the gene was protein or DNA. (Actually, he may have been somewhat biased toward the protein hypothesis, but so was everyone else in the early 1940s.) Yet he saw in the pneumonia bacterium a way of isolating the genetic structure and demonstrating its nature once and for all.

To this end, Avery battered the dead pneumonia bacteria with an arsenal of biological weapons designed to purify their various substances so that each could be studied in isolation. When Avery exposed the harmless, rough-coated bacteria to protein extracted from the dead, virulent bacteria, nothing happened. But when he exposed the harmless bacteria to DNA extracted from the virulent bacteria, the former suddenly gained a smooth coat and became virulent itself. Some kind of genetic message had been passed on. And it had been passed on by DNA.

The weight of opinion that held that genes were made of protein was great, however, and Avery's experiment, while it made a dent in that body of opinion, did not overwhelm it. Avery, a methodical researcher who insisted that his findings be backed up by exhaustive experimentation, announced his results in 1944. It was nearly a decade after that before the gene was shown conclusively to reside in DNA.

That task fell to James Watson and Francis Crick, researchers at Cambridge University in England (though Watson is an American), who took X-ray photographs of DNA and built painstaking models of the molecules based on the ways in which they scattered the X rays onto the photographic plate. Yet the key insight that allowed Watson and Crick to elucidate the nature of the gene came less from experimental evidence than from an inspired guess as to the mechanism that allowed chromosomes to make copies of themselves. That mechanism is sometimes called *base pairing*. What makes the base pairing hypothesis so marvelous is that it not only explains the copying mechanism of the chromosome, but it also explains how information is transcribed in the genes and how that information is used to control the production of enzymes.

THE MODEL MOLECULE

The picture of DNA elucidated by Watson and Crick (and the others whose work they built on) was essentially this:

The average DNA molecule is a long polymer strung together from smaller molecules called nucleotides. The nucleotides are in turn made up of three even smaller molecules—a *sugar molecule*, a *phosphate molecule*, and a third type of molecule called a *base*.

The sugar and phosphate molecules are purely structural—that is, they have nothing directly to do with chromosome replication or with carrying genetic information. Endlessly repeated along the length of the chromosome, the sugars and phosphates form the *molecular spine*, a monotonous chain of atoms that holds the chromosome together. The bases, in turn, stick out at right angles from this spine like leaves protruding from a vine or branches from a tree trunk. There are four different kinds of bases in DNA: *adenine*, *guanine*, *thymine*, and *cytosine*. Each has a slightly different structure. (In RNA, a fifth base, *uracil*, is substituted for thymine. Apparently this is nature's way of "tagging" the RNA molecule so that it will not be confused with DNA.)

These bases are the carriers of genetic information. Like letters in a word or words in a sentence, they spell out recipes for constructing enzymes, according to a system known as the *genetic code*. How this system works we will see in a moment; first, let's examine the way in which the chromosomes duplicate themselves.

Is the DNA molecule, then, only a single chain of phosphate and sugar molecules, with bases dangling from it at intervals? No—

actually DNA is a double chain of phosphate and sugar molecules, each with its own set of bases. You might imagine the DNA molecule as two parallel chains of nucleotides; the bases on one chain reach out and bond with the corresponding bases on the parallel chain, linking the two chains together. We might compare this double chain to an old-fashioned ladder: The sugar-phosphate spines are the legs of the ladder, and the linked bases are the rungs, holding the legs together. (Each rung in this analogy represents two bases—one extending from each spine, both meeting in the middle.)

Yet even this image is not quite accurate. The "legs" of the DNA ladder are not straight and rigid; they are twisted into a coil, a helix, a sort of molecular spiral staircase. The two helixes—which together form the well-known "double helix" of DNA—wind endlessly around one another, like the intertwined serpents on a physician's caduceus. These two coils are connected by the linked bases.

This is where base pairing comes in. Chromosome duplication is based on a deceptively simple, natural circumstance: Not every

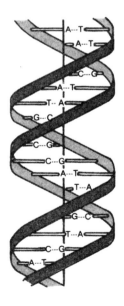
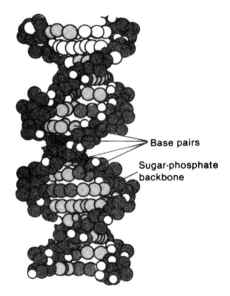

Base pairs

Sugar-phosphate backbone

In this diagram, you can see that the DNA molecule is a double helix—literally, doubled spiral coils—composed of two chains. The dark outer ribbons are sugar-phosphate backbones that twist around, while the paired bases (adenine/thymine, guanine/cytosine) on the inside hold the chains together.

base will bond with every other base. A base on one strand of the double helix will reach out and clasp the base on the other strand only if it is the right type of base. Adenine will attach itself only to thymine (or to uracil, in RNA) and guanine will attach itself only to cytosine. Other bonds simply aren't possible: The shapes of the molecules and the locations of their bonds are not compatible. Like two mismatched pieces in a jigsaw puzzle, they refuse to fit together. If cytosine, to choose a single example, tried to attach itself to thymine, the bond would not take. The bases would fall apart as quickly as they came together. Only when cytosine encounters guanine will an attachment form.

Consider what this means. If we find an adenine base on one strand of a DNA molecule, we know that the base to which it is bonded on the opposite strand must be thymine. No other combination will work. If we find a cytosine base on one strand, we know that it will be bonded to a guanine base at the corresponding point on the opposite strand. While the order of bases on one strand may seem to us quite random (though it actually is not, no more than the words and letters in this sentence are random), once we know the order of bases on one strand, the order on the other strand becomes quite predictable. The two strands are, in a sense, negative images of one another, with complementary arrangements of bases. For instance, if one strand reads "adenine–guanine–guanine–cytosine–adenine–thymine" (which we may abbreviate as A–G–G–C–A–T), we know immediately that the corresponding segment of the parallel strand will read "thymine–cytosine–cytosine–guanine–thymine–adenine" (or T–C–C–G–T–A).

When the chromosome is ready to duplicate itself, an enzyme "unzips" the double helix, breaking the weak hydrogen bonds by which the bases on one strand are attached to the bases on the other. As the bonds break, the two strands of the DNA molecule disengage. They float away from one another, the double helix now broken into two single strands of nucleotides.

The bases on the nucleotides now are left naked, exposed, ready to bond with other bases. Fortunately, the nucleus of the cell is full of loose nucleotides, each with a base waiting to be bonded and a sugar-phosphate combination ready to become part of a chromosome spine. Yet each of these bases will bond only with its complementary base—guanine to cytosine, thymine to adenine.

In a process abetted by several different enzymes, these loose nucleotides are thrown against the naked bases of the detached

DNA strands. If the correct bases come together, a bond will form. If not, the loose nucleotide will bounce away unbonded; other nucleotides will attempt to attach themselves to the exposed base on the chromosome until the right one is found. For instance, if there is a guanine base on the chromosome, it will bond only with a cytosine base on a loose nucleotide. Nucleotides with other bases may attempt to bond with the guanine, but they will not succeed. A nucleotide with a cytosine base, however, will be accepted immediately; in such fashion do all the exposed bases on the chromosome pair off with their complementary bases on the loose nucleotides.

As base attaches to base, thereby securing the loose nucleotides to the DNA strand, the phosphate-sugar portions of these nucleotides click into place to form a new helical spine intertwined with the spine of the existing strand. Thus a second strand forms twisted around the first. Within a few minutes, the single strand of DNA has become, once again, a double helix, the second helix formed out of loose nucleotides from within the nucleus.

In fact, because this process takes place along both of the single strands created by the splitting of the original double helix, we now have two double helixes where there was one before. The chromosome has duplicated itself.

It is important to note that both of these new double helixes are identical to the original. For instance, suppose that a segment of the original double strand looked like this (where A, T, G, and C stand for the bases attached to the nucleotides):

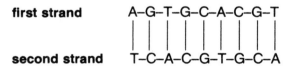

first strand   A–G–T–G–C–A–C–G–T

second strand   T–C–A–C–G–T–G–C–A

Note that the nucleotides in one strand are paired with nucleotides in the other strand according to the strict rules of base pairing (A to T, T to A, C to G, and G to C). When this double strand divides into two single strands, the resulting strands will look like this:

A–G–T–G–C–A–C–G–T

and

T–C–A–C–G–T–G–C–A

As the bases bond with their corresponding bases, out of the nearly inexhaustible fund of nucleotides floating around loose within the nucleus of the cell, they eventually form two double strands that look like this:

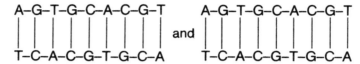

Each of these strands is, of course, the same as the original; they could hardly be otherwise. These double strands then can split to form still more strands, allowing the chromosomes to make as many copies of themselves as may prove necessary.

In this manner, the coded information on the chromosomes can be duplicated as many times as needed, so that each cell in your body can have its own copy.

It's only fair to add that this description simplifies the process considerably. For instance, chromosome duplication actually is a rather sloppy affair, enough so that the cell contains a class of enzymes designed solely for proofreading genetic information. Such enzymes examine the twin strands of the double helix for possible mismatching of nucleotides, and when such a mismatch is found, still other enzymes are called into action and the strand is edited, the misreading corrected. In some instances, the enzymes can even correct mistakes that leave both strands hopelessly garbled, by comparing the genetic information on one molecule with information on other molecules of DNA.

When chromosome damage occurs at a high rate—as when the cell is exposed to an onslaught of radiation—quick and dirty repairs can be performed on the DNA strand, with undamaged nucleotides being hammered into place at random. Once the emergency has passed, these random nucleotides can be replaced with the proper sequence of nucleotides, assuming another molecule containing that sequence is available for comparison. If no such comparison is available, the random sequence of nucleotides remains and is copied every time the chromosome duplicates itself. If this damage occurs in a significant gene, the garbled genetic information can cause the death of the cell. If it occurs in a sperm or egg cell, this genetic damage may be passed on to the next generation (assuming that the egg cell survives the damage). Such genetic aberrations are called

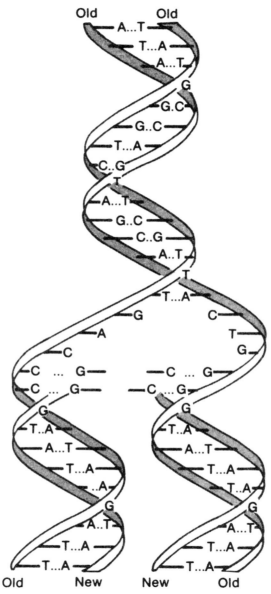

When DNA replicates, the original strands unwind and serve as templates for the building of two new complementary strands. The two new strands of DNA are called *daughter molecules;* the original is called the *parent.* The two daughter molecules are exact copies of the parent because each of the two has one of the unraveled parent strands.

mutations; they are one of the driving forces behind the slow process of genetic change known as evolution.

CONCEPTUAL QUESTIONS

The very concept of the gene raises a slew of tantalizing questions.

How is genetic information carried in DNA molecules? In what language is it written? How is it transcribed? How is this information translated into chemical form?

DNA, you will recall, is only one of two forms of nucleic acid; the other is RNA. While DNA is found almost entirely in the chromosome, RNA is distributed throughout the cell. And while DNA carries the genetic information, RNA reads it.

Both RNA and DNA are constructed from nucleotides, but the nucleotides of RNA are different from those of DNA in two essential ways: (1) The sugar molecules in the "spine" are of slightly different construction, an arrangement that may prevent loose RNA nucleotides from accidentally becoming incorporated in the chromosomes; and (2) the base uracil is substituted for the base thymine. (Like thymine, however, uracil will bond with adenine—and only with adenine.) Furthermore, there are three different kinds of RNA, two of which play essential roles in enzyme production—*messenger RNA* (abbreviated *mRNA*) and *transfer RNA* (*tRNA*).

As with DNA, loose nucleotides of mRNA float freely in the nucleus of the cell, buffeted by the heat motion of the cell's interior. When selected genetic information is required for the production of organic molecules elsewhere in the cell, enzymes unzip a portion of the double helix—the portion containing the needed information and no other—and allow the loose mRNA nucleotides to attach themselves to the exposed nucleotides of the DNA molecule. RNA attaches itself to DNA in roughly the same way that DNA attaches to itself—guanine to cytosine, uracil (instead of thymine) to adenine, cytosine to guanine, and adenine to thymine. In this way, the RNA makes a "copy" of the sequence of nucleotides on the DNA and thereby copies the genetic information contained in that sequence. This process is called *RNA transcription*.

For instance, consider the imaginary strand of DNA that we examined earlier:

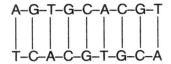

As the chromosome is unzipped for RNA transcription, only one strand would be exposed to the bombardment of RNA nucleotides; the other strand contains no useful information but exists only for duplication purposes. Nucleotides of mRNA would attach themselves to the exposed strand as follows:

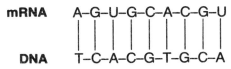

Note that the correspondence between parallel nucleotides is identical with that in the double helix, except that uracil has been substituted in the RNA strand.

At this point the mRNA strand breaks loose from the DNA molecule and floats away in isolation:

A–G–U–G–C–A–C–G–U

In fact, it flees the nucleus altogether and searches out the nearest ribosome.

The ribosome, in a sense, is a machine for reading the coded message on the mRNA and for transforming that message into actual molecules. In this it is abetted by the second kind of RNA: transfer RNA. To see how this is done, we have to examine the nature of the genetic code, the language in which genetic information is written.

LANGUAGE OF THE CELL

As we saw a moment ago, the genetic code is expressed in the order of the nucleotide bases. Like the words and letters in the sentence you are reading now, the bases spell out a message which the ribosomes, in conjunction with tRNA, can "read." The message consists, simply, of a list of amino acids which, when joined together in sequence, will comprise an enzyme.

To read this message, we must consider the bases on the DNA molecule (or, in this case, the mRNA copy of that molecule) as occurring in sets of three. A set of three bases is called a *triplet* (or, sometimes, a *codon*). Broken up into triplets, our imaginary mRNA strand would look like this:

AGU–GCA–CGU

Because there are four different bases in the genetic alphabet, the maximum possible number of such triplets would be $4 \times 4 \times 4$, or 64. Each of these 64 triplets, including the three shown above, stands for a specific amino acid. Checking the chart on page 23, we see that the three triplets in our imaginary molecule stand for these amino acids:

serine–alanine–arginine

Therefore, our mRNA molecule represents an enzyme (or segment of an enzyme—three amino acids is a very short molecule) containing those amino acids in that order.

(It should be noted here that there are triplets that do not stand for amino acids; some are codons identifying the termination of a sequence of amino acids, as you can see on the chart. Furthermore, because there are sixty-four possible codons and only twenty amino acids, each amino acid is represented by several different triplets. This redundancy factor decreases the risk inherent in the accidental miscopying of genetic information during the duplication of DNA.)

The codons are translated into amino acids through the intervention of transfer RNA. Each tRNA molecule carries on it three nucleotides—no more, no less; the nucleotides are located at one end of the molecule, like a rubber stamp attached to a handle. These nucleotides, taken together, form an RNA triplet, the specific triplet varying from molecule to molecule. For each possible triplet of mRNA there exists a molecule of tRNA with a corresponding triplet spelled out on its attached nucleotides. The triplets of tRNA are not identical to the triplets of mRNA; they are complementary, according to the rules of base pairing. For instance, the tRNA triplet corresponding to the mRNA triplet AGU would be UCA. This allows the tRNA triplet to attach itself to the complementary triplet on the mRNA molecule, with A bonding to U, G to C, and U to A.

Why would tRNA bond with mRNA? At the other end of each tRNA molecule, opposite the triplet, is an attached amino acid—the specific amino acid coded for by the triplet on that molecule (or, rather, its corresponding mRNA triplet). If, for instance, the triplet on the tRNA molecule is UCA (which corresponds to the mRNA triplet AGU), the amino acid attached to that molecule would be serine. And, checking the chart on page 23, we see that the codon representing serine is AGU.

We might envision the tRNA molecule as looking something like this:

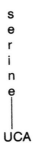

The mRNA strand, once it escapes from the nucleus, attaches itself to a ribosome. It passes through this cell structure much as a length of ticker tape passes through a teleprinter, all the while being bombarded by the tRNA molecules that float loosely in the outer part of the cell. This bombardment is random (though facilitated by enzymes); most tRNA molecules simply bounce away. However, when the triplet attached to a loose tRNA molecule encounters its corresponding triplet on the mRNA strand, it will attach itself. Momentarily, a sequence of tRNA molecules will form along the strand, rather like this:

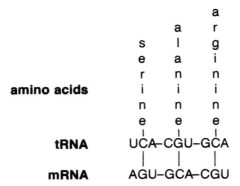

At a certain point in their passage through the ribosome, the amino acids—until now oriented at right angles to the RNA strand—are repositioned and attached to one another, creating the following protein molecule:

serine–alanine–arginine

As they leave the ribosome, the two kinds of RNA and the chain of amino acids go their separate ways. The amino acid chain takes on a distinct shape: An amino acid at one point on the chain may form a bond with another amino acid considerably farther along the chain; such bonds cause the strand to double back on itself, thus giving the enzyme its globular shape.

Once the enzyme is released into the cell, it immediately begins promoting chemical reactions. The activity of the cell—and therefore of life itself—is perpetuated.

THE GENETIC CODE

Nucleotides	Amino Acids
UUU	phenylalanine
UUC	phenylalanine
UUA	leucine
UUG	leucine
UCU	serine
UCC	serine
UCA	serine
UCG	serine
UAU	tyrosine
UAC	tyrosine
UAA	stop
UAG	stop
UGU	cysteine
UGC	cysteine
UGA	stop
UGG	tryptophan
CUU	leucine
CUC	leucine
CUA	leucine
CUG	leucine
CCU	proline
CCC	proline
CCA	proline
CCG	proline
CAU	histidine
CAC	histidine
CAA	glutamine
CAG	glutamine
CGU	arginine
CGC	arginine
CGA	arginine
CGG	arginine

Nucleotides	Amino Acids
AUU	isoleucine
AUC	isoleucine
AUA	isoleucine
AUG	methionine
ACU	threonine
ACC	threonine
ACA	threonine
ACG	threonine
AAU	asparagine
AAC	asparagine
AAA	lysine
AAG	lysine
AGU	serine
AGC	serine
AGA	arginine
AGG	arginine
GUU	valine
GUC	valine
GUA	valine
GUG	valine
GCU	alanine
GCC	alanine
GCA	alanine
GCG	alanine
GAU	aspartic acid
GAC	aspartic acid
GAA	glutamic acid
GAG	glutamic acid
GGU	glycine
GGC	glycine
GGA	glycine
GGG	glycine

THE LITTLEST LIBRARY

In every cell of every human body there are thirty-four pairs of chromosomes, each containing tens of thousands of genes. (In modern biological terminology the gene is that portion of the chromosome that codes for a single enzyme molecule—or, in the case of enzymes constructed out of more than one molecule, for a single molecular unit of that enzyme.)

What a fabulous store of information the chromosomes represent! Within this vast library are the "recipes" for every enzyme that the organism as a whole might possibly need, tens of thousands of them. But who, or what, decides when these recipes are to be used? Information, no matter how detailed, is useless without some provision for discriminating between the necessary and the unnecessary. Although there are a few enzymes that are needed almost constantly to perform the ordinary housekeeping functions of the cell, most are required only on special occasions. When these occasions arise, what sets the molecular machinery in motion?

This question was addressed in the late 1950s by French biologists Jacques Monod and François Jacob of the Institut Louis Pasteur in Paris. The theory of genetic control that they devised came to be known as the *operon theory*; it hinges on a process called *negative feedback*, which says, in essence, that the cell manufactures any enzyme it has not been explicitly forbidden to manufacture.

Related genes tend to be located next to one another on the chromosome—that is, when more than one enzyme is needed to perform a chemical task, the genes for those enzymes usually are found in sequence, one after another. (There are, it should be pointed out, some notable exceptions to this rule.)

Why should this be?

Monod and Jacob called these groupings *operons* and theorized that they existed to facilitate the control of enzyme production. When the information on a gene is transcribed onto mRNA molecules (the first step in enzyme production, as we saw earlier), there are special enzymes which aid and abet in the task. These enzymes attach themselves at a point on the chromosome just above the operon containing the genes they wish to transcribe. The point at which this attachment takes place is called the *operator gene*. In order for transcription to take place, the enzyme must be attached to the operator.

It follows logically, then, that if this enzyme can be prevented from attaching to the operator, the genes in the operon will not be transcribed into mRNA. And if mRNA copies of the genes cannot be made, no enzymes can be produced.

This is how the cell controls the production of enzymes. For each operator gene on the chromosome, there is a *repressor gene* located elsewhere in the nucleus of the cell. The repressor gene produces a *repressor molecule*, which can attach itself to the operator gene in such a way that it blocks the mRNA transcription enzyme, preventing it from performing its function. In this way the cell "forbids" the production of an enzyme or sequence of enzymes.

What happens when those enzymes are required by the cell? How is the operon freed for transcription?

Monod and Jacob studied this problem in bacteria, observing the activity surrounding the gene for beta galactosidase, an enzyme that breaks down galactose sugar molecules into their component parts. This enzyme is required only when this sugar is present in the cell; at other times it is blocked off by a repressor molecule.

Through a series of elegant experiments, Monod and Jacob deduced that the galactose sugar molecule is itself responsible for removing the repressor molecule from the beta galactosidase operon. When the sugar molecules flood the cell, they interact physically with the beta galactosidase repressor, pulling it loose from the operator gene, allowing the transcription enzyme to attach itself and transcribe the operon. The beta galactosidase enzymes are produced and the galactose molecules are broken down.

Once the sugars have been dismantled, however, the repressor molecules are free to attach themselves once more to the operator gene. Production of the enzyme comes to a halt.

The marvelous part of this theory of genetic control is that it is entirely self-contained. The presence of a molecule in the cell calls up the enzymes needed to deal with that presence. And the absence of the molecule banishes the enzyme to limbo.

As ingenious as the operon theory is, it leaves a number of questions unanswered. How, for instance, does one cell differentiate itself from other cells? A skin cell has very different enzymatic requirements from a brain cell. The chemical reactions promoted by a lung cell are not those of a liver cell. Yet each of these cells contains the same set of chromosomes, the same set of genes, and supposedly the same operator-repressor system. What sets these various kinds of cells apart? This, alas, is one of the great unanswered questions of biology; only time and diligent research will provide an answer.

THE MAN-MACHINE

Perhaps the most remarkable thing about these remarkable processes is their almost mechanical nature. Like some Rube

Goldbergian contraption thrown together out of old hubcaps and lollipop sticks, the molecular mechanisms of the cell seem almost antithetical to what we think of when we think of life. Life is soft, warm, infinitely changeable, infinitely mysterious. The processes by which DNA governs the action of the cell seem surprisingly rigid and precise, altogether machinelike.

There is little room for vitalism in molecular biology. The rules of physics apply as much to genes as to billiard balls and ballistic missiles. Pasteur to the contrary, there is no significant difference between life and nonlife except one of complexity. The organic obeys the same laws as the inorganic.

Consider what this means. If the essence of life is a mechanical process of information exchange whereby the coded genetic information is translated into the chemicals needed by the cell, it should be possible, in theory at least, to edit that information, alter it, change its flow, and thereby change the nature of the organism—in short, to engineer a living organism as a mechanic engineers a custom race car.

But how would we go about this? Biologists have developed microsurgical techniques so delicate that they can perform surgical operations on living cells, but it is difficult to conceive of any man-made surgical tools tiny and precise enough to reconstruct a molecule, even a molecule as large as the chromosome. Yet, might it not be possible to use tools that are not man-made but that already exist in nature, tools created by the genes themselves? Earlier, we described enzymes as molecular machines. What if we could turn those machines to our own ends, harness them for the manipulation of other molecules?

That, in fact, is the essence of genetic engineering—the use of the mechanisms of life to alter the mechanisms of life. And it is the key to the genetic revolution.

PART TWO

Shaping the Gene

DESIGNER GENES

If genetic engineering is possible, has it been done? And if it has, when? What was the first act of deliberate genetic engineering performed by a human being?

This is more a matter of definition than of history. The first person to breed food crops by cultivating desirable plants in preference to undesirable plants was performing a crude genetic manipulation. By selecting those species with "good" genes over those species with "bad" genes, the genetic composition of a species can be profoundly changed. Certainly there are thousands of plants and animals on earth today that never would have existed without this kind of artificial selection.

But the sort of genetic engineering under discussion here is a far more intimate process, one that enters the very nucleus of the cell and reshapes the chromosomes. This sort of molecular tinkering is very much a twentieth-century accomplishment, as far from conventional agricultural practices as the automobile is from the rickshaw.

Nobel laureate René Dubos has argued that Oswald Avery performed the first act of genetic engineering when he turned harmless bacteria into virulent bacteria by exposing them to an extract of DNA. But Avery simply transplanted a gene from one individual of a species to another; no alteration or unusual recombination of genetic material took place.

Others might argue that genetic engineering began with the work of Severo Ochoa and Arthur Kornberg. Ochoa, an American biochemist born in Spain, discovered in the early 1950s a bacterial enzyme that promoted the formation of RNA chains. When added to a solution of loose RNA nucleotides, the enzyme encouraged those nucleotides to link together into a polymer—and thus an "artificial" molecule of sorts was formed. But there was no way to specify what genetic information was carried on the strand; the process was totally spontaneous.

Working roughly at the same time as Ochoa, Kornberg sought an equivalent enzyme for DNA. He found it in an extract from the bacterium *Escherichia coli* (*E. coli*, for short). By June 1956 he had purified that extract down into a pure enzyme, which he termed DNA polymerase, because it created DNA polymers. The enzyme, Kornberg discovered, would assemble new DNA molecules, but only in the presence of an old DNA molecule, which could serve as a template for DNA replication. The new molecules were close replicas of the originals but not precise enough for normal organic purposes. It was twelve years before Kornberg developed a process that would produce biologically active DNA in the test tube.

Could the work of either Kornberg or Ochoa be termed genetic engineering? Not exactly. They created genetic materials but could not alter those materials; no meaningful form was given to the molecules that had not existed before. And no attempt was made to place these new molecules within living systems. This was not to happen until the 1970s, with the arrival of three scientists named Paul Berg, Stanley Cohen, and Herbert Boyer.

THE TOOLS

The line between life and nonlife is a vague one. Viruses fall in a gray region that is not quite either.

If viruses are genuinely alive, they are certainly the simplest of organisms—no more than a shell of protein surrounding a coil of DNA (although some viruses contain RNA rather than DNA). Viral DNA contains all of the instructions necessary for making new viruses identical to the parent organism, but the virus has no facilities for making those viruses itself—and without reproductive machinery, its DNA is useless.

Of necessity, then, the virus is forced to hijack the reproductive machinery of living cells. Like a miniature hypodermic needle, the virus attaches itself to the outside of a cell and injects its DNA into the cytoplasm—the substance of the cells. Within the cell, the DNA begins to manufacture the proteins it needs to create new viruses, and soon the cell is filled to bursting with the viral progeny. The cell erupts and the newborn viruses pour outward in search of new cells to infiltrate. Alternatively, the viral DNA can physically insert itself into the cell's chromosome, becoming part of the organism's genetic material. (If the virus carries its genetic information on RNA rather than DNA, it first must make a DNA copy of the RNA, a process called reverse transcription, abetted—appropriately enough—by an enzyme called reverse transcriptase.) When the time comes for the virus to reproduce, the DNA produces enzymes that excise it from the chromosome so that it can initiate the reproductive process.

Although the virus has nothing in particular against the cell—it only needs a convenient place to reproduce its kind—the cell often gets the worse of the relationship. Consequently, most cells have developed weapons for repelling foreign DNA. Some of these weapons fall under the general heading of *restriction enzymes*—tiny molecular scissors that slice undesirable DNA molecules into shreds before they can cause damage.

Though no metallurgist could forge tools this small, these restriction enzymes can be used by molecular biologists as instruments for the study and manipulation of chromosomes. One of the first to utilize these tools in laboratory research was Paul Berg of Stanford University.

In 1971, Berg was engaged in the study of cell differentiation and the ways in which gene expression was controlled within the cell. Berg decided that it would be valuable to remove specific genes from the chromosomes of certain organisms and splice those genes into other organisms, where they could be studied in isolation. If the gene expressed itself in its new host (or even if it did not), Berg reasoned that he would learn a great deal about the ways in which that expression was controlled.

To that end, Berg devised an elegant experiment. He would expose the DNA of a virus called SV40 (which causes cancer in certain species of animals) to a restriction enzyme. The enzyme would break the DNA into fragments, which Berg could insert onto the cells of bacteria, where they could be studied.

But how could Berg get the DNA fragment into the bacterial cell? His solution to this problem was ingenious. By using an enzymatic process not unlike that employed by Kornberg in synthesizing DNA, Berg added DNA "tails" to both ends of the fragment.

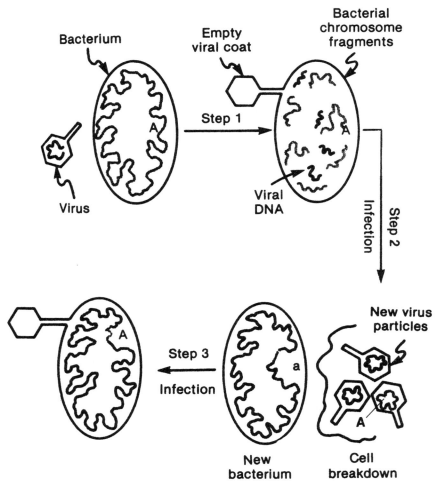

Viruses have been called nature's hypodermic needle; evidently they've been tampering with DNA for eons. Viruses literally inject their own DNA into a cell in order to reproduce. In the illustration above, a virus comes along and injects a cell with its DNA. When the infected cell breaks down, several new virus particles are produced. However, as sometimes happens, one of them is carrying chromosomes from the host cell *(A)*. When this virus injects its DNA into a cell, the original host cell's genetic element can recombine and replace a similar segment *(a)* in the new host, thus exchanging gene *A* for gene *a*.

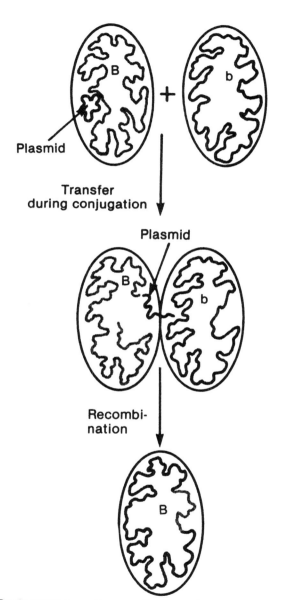

Conjugation is another way that genetic materials recombine naturally. In conjugation, a plasmid in one bacterium (left) can transfer a bacterial chromosome *(B)* to a second cell, causing the exchange of gene *B* from the first cell for gene *b* in the second cell.

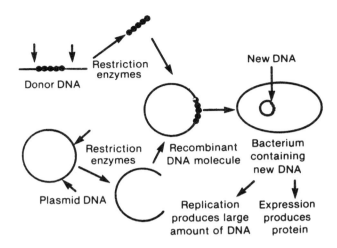

Donor DNA

Restriction enzymes

New DNA

Restriction enzymes

Plasmid DNA

Recombinant DNA molecule

Bacterium containing new DNA

Replication produces large amount of DNA

Expression produces protein

Restriction enzymes recognize certain sites along the DNA molecule and can chemically cut the molecule at those sites. This makes it possible to take specific genes from donor DNA molecules and insert them into other DNA molecules, forming something new and different—recombinant DNA. This re- combinant DNA can then be cloned, and large amounts of a desired substance can be produced. *(courtesy Office of Technology Assessment)*

Each tail consisted of a single strand of DNA rather than a double strand; because this left the bases exposed, the tails were free to bond to other single strands of DNA with complementary bases, according to the rules of base pairing. In effect, Berg had given the DNA "sticky ends," which could easily be coerced into attaching themselves to another set of complementary ends.

He then obtained more DNA, this time from a type of virus called a *bacteriophage*, which preys exclusively on bacteria. (The term bacteriophage literally means "bacteria-eater.") Berg broke apart the bacteriophage DNA with restriction enzymes and attached similar sticky ends to one of the fragments, grafting on tails that were precisely complementary to those on the SV40 DNA. Then, with the aid of enzymes that normally repaired breaks in DNA molecules, he attached the sticky ends of the bacteriophage DNA to the sticky ends of the SV40 DNA. The two molecules joined together to form a single molecule, which he then inserted into the protein coat of the bacteriophage.

It would have been simple enough for Berg to get this hybrid

Paul Berg was the first to utilize the genetic engineer's best tool—restriction enzymes. He also was the man the scientific community turned to when questions arose about the safety of these new procedures. He chaired the first Asilomar Conference. *(courtesy National Academy of Sciences)*

DNA inside a bacterial cell; a bacteriophage is designed by nature to inject its DNA into bacteria. But this was an experiment that Berg never undertook, for reasons we will discuss shortly.

In a sense, Berg had opened the door to genetic engineering but had refused to step through. His hybrid DNA molecule represented a fusion of the genetic material of two different organisms—two different species of organism, in fact—but it was not alive. It would not be alive until it was introduced into a living organism, where it could become part of that organism's genetic contents.

That particular step was left to Cohen and Boyer two years later.

BREAKING THROUGH

Living cells can be divided into two types: *eukaryotic* and *prokaryotic*. The former keep their chromosomes locked away behind the protective walls of the nucleus (and therefore are sometimes referred to as nucleated cells). In the latter, the chromosomes float around loosely in the cytoplasm, mixing with the various organelles. Prokaryotic cells tend to be comparatively primitive, more advanced perhaps than the first cells to emerge from the primordial oceans, but not capable of forming the complex multicellular organisms of which eukaryotic cells often are a part. Prokaryotic cells are, in fact, invariably single-celled organisms; only eukaryotic cells can join together to create higher forms of life.

Bacteria are prokaryotic. The single chromosome of the bacterial cell (a long double helix of DNA in the form of a loop) floats unhindered through the body of the cell, but it does not always float alone. Coexisting with the normal chromosome loop of the cell may be other, smaller loops of DNA. These loops, called *plasmids*, are like benevolent viruses—in a sense, they are separate organisms made up entirely of DNA, existing if not exactly living in a symbiotic relationship with the bacterium. What is the nature of this symbiosis? The bacterium gives the plasmid a warm place to live and the use of its protein synthesizing equipment. The plasmid, in turn, gives the bacterium some rather valuable genes.

Of what use are plasmid genes to a bacterium? Like a house guest bearing expensive gifts to make himself welcome, the plasmid may carry genes for such valuable traits as antibiotic resistance. A bacterium carrying such a plasmid might become resistant to penicillin or to tetracycline. Bacteria willing to carry plasmids as excess baggage therefore are conferred with a strong advantage in the evolutionary sweepstakes.

So welcome have these plasmids made themselves in the world of prokaryotic cells that bacteria tend to pass them around among themselves, like children trading baseball cards. A single plasmid in a bacterial cell can replicate itself much as the bacterium's own chromosome replicates itself (though there are certain differences in the mechanics of the process). Not requiring an extra copy of the plasmid, the bacterium can transfer it to another bacterium through a process called conjugation, which vaguely resembles the sexual activity of higher organisms (and which exists for much the same purpose—the spreading of desirable and varying genetic traits through a large sector of the population). This presents certain problems to those of us for whom bacteria sometimes are less than welcome parasites. A plasmid that confers resistance to penicillin, say, will render a virulent (disease-causing) bacterium that much more dangerous. And because doctors tend sometimes to overprescribe antibiotics such as penicillin to patients with mild bacterial infections, there tends to be a strong selective pressure in favor of bacteria carrying these plasmids, which in turn increases the number of the plasmids in circulation. Ordinarily, this selective pressure works hardest on the benevolent bacteria that inhabit the human intestines, but these bacteria then can pass on their plasmids to any more virulent bacteria they may come in contact with. Thus strains of antibiotic-resistant bacteria have become increasingly prevalent in recent years.

Be that as it may, plasmids have proven unexpectedly useful to molecular biologists as vehicles for carrying hybrid DNA molecules into bacterial cells.

Stanley Cohen was, and remains, an expert on plasmids.

The antibiotic-resistant properties of plasmids were first recognized in the mid-1960s and Cohen, a Stanford University researcher, shortly joined the search for what had come to be called the R factor—the plasmid gene that codes for antiobiotic resistance.

To isolate the genes of specific plasmid molecules, Cohen whirled them about in high-speed laboratory blenders, battering

Herbert Boyer was—and is—an enzyme expert. His work with
Stanley Cohen started the age of genetic engineering and also
led to Boyer's starting the very successful company Genentech. *(courtesy National Academy of Sciences)*

them into fragments which could be studied separately in the test
tube; sometimes he allowed these fragments to join back together
into altered loops of DNA. The method was unsatisfactory, but it was
all that was available.

Not so for long. In 1972, Cohen met Herbert Boyer, a biologist
from the University of California at San Francisco, at an international conference in Hawaii. Whereas Cohen's fascination was with
plasmids, Boyer's was with enzymes, specifically restriction enzymes. Over a late dinner in a Waikiki delicatessen, Boyer described the marvelous abilities of these tiny molecular instruments.
Not only did they sever DNA molecules cleanly and efficiently,
Boyer related, but, in addition, they did so only at certain points on
the chromosome. Each restriction enzyme was designed by nature
to "recognize" a specific sequence of nucleotides and would cut into
the chromosome only at that sequence; these enzymes, therefore,
could be used in a precise and predictable manner. Because different restriction enzymes cut DNA at different nucleotide sequences, they offered the biologist a choice of sites at which to edit
the molecule.

As though this were not enough, one of these restriction enzymes, dubbed EcoRI, even produced its own sticky ends. Whereas Berg had gone through the painstaking process of linking complementary tails to his hybrid molecules, EcoRI left such tails as a matter of course, shearing one strand of the double helix at a point several nucleotides downstream from where it had sheared the complementary strand, leaving the intervening bases over that distance exposed and unpaired. And because every DNA molecule cut by EcoRI was severed at the same sequence of bases, the tails of each fragment were naturally complementary to one another.

This process, central to the concept of recombinant DNA, bears closer examination. The sequence of bases where EcoRI cuts the DNA molecule is this:

$$G-A-A-T-T-C$$
$$| \quad | \quad | \quad | \quad | \quad |$$
$$C-T-T-A-A-G$$

The restriction enzyme wields its scalpel at one end of this sequence on the upper strand and at the opposite end on the lower strand. (Interestingly, and by no coincidence, the two sequences—the upper and the lower—are mirror images of one another, so that EcoRI actually is clipping at the same point on the sequence on each strand, but at a slightly different location on the double helix.) The nucleotides between the two points are unzipped by the enzyme, so that the bases are exposed. The result is that each fragment of the molecule has a tail of nucleotides with exposed bases. The tail on one fragment looks like this:

$$G-A-A-T-T-C$$

And the tail on the other fragment looks like this:

$$C-T-T-A-A-G$$

Clearly these tails are complementary to one another, but they are also complementary to the tails on any other molecules severed by EcoRI. If two plasmids, say, were subjected to the restriction enzyme and if the resulting fragments were thrown together into a test tube, they could conceivably, with the proper enzymatic assistance, recombine into a single plasmid loop by joining their complementary ends.

Cohen saw this clearly as Boyer described his research. EcoRI represented an ideal tool for dissecting plasmids and exploring their varied parts. Boyer, in turn, recognized that Cohen's plasmids offered an instrument nearly as valuable—a vehicle for inserting fragmented molecules into the cells of bacteria.

And such it proved to be. Cohen and Boyer returned to the mainland as scientific collaborators and soon discovered among Cohen's collection of battered and reconstructed plasmids a tiny loop of DNA that contained precisely one site at which it could be cut by EcoRI—that is, the sequence of nucleotides recognized by the restriction enzyme (G–A–A–T–T–C) occurred at only one spot on the plasmid.

Thus the plasmid could be converted from a loop of DNA into a linear strand of DNA with sticky ends, through the application of EcoRI. If a second piece of DNA from another organism also were exposed to EcoRI, the fragments of that molecule would all have sticky ends complementary to those in the broken plasmid. And if the fragments of both molecules were placed together in a test tube along with the proper enzymes, they would join their sticky ends together and form a single, continuous loop—a recombinant plasmid, as it were—which could be inserted into a bacterium, where the genes might be induced to express themselves.

The plasmid with the single excision site was dubbed pSC101—for "plasmid Stanley Cohen 101." In 1973, Cohen and Boyer spliced into one of these plasmids genes taken from a staphylococcus bacterium and genes taken from a toad. Then they inserted this recombined plasmid into the bacterium *E. coli*, a common microorganism ordinarily found in the human intestines. Once in the bacterium, certain genes in the plasmid expressed themselves.

It was an historic moment. For the first time, genes from several different species had been placed in a single organism in such a way that they became part of the working genetic complement of that organism, forming a kind of hybrid species.

The age of genetic engineering had begun.

AFTER THE BEGINNING

Why would anyone want to splice genes into a bacterium?

There are quite a few reasons, not the least of which was to see if it could be done. More importantly, it offered the biologist a means

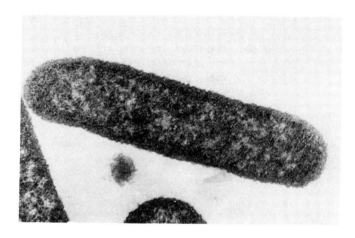

This is an *Escherichia coli* bacterium magnified 60,000 times. Most recombinant-DNA experiments use *E. coli* as host cells because it reproduces rapidly and thrives under laboratory conditions. *(courtesy National Institute of Allergy and Infectious Diseases)*

of studying genes in isolation. Interesting segments of DNA could be removed from a chromosome and inserted into another kind of organism. If the new host then should begin producing an enzyme it had never produced before, very likely that enzyme was the one coded for by the inserted gene—and thus the function of that gene, and perhaps even the way in which that function is controlled, is revealed. If a biologist fragments an entire chromosome and inserts those fragments at random into bacteria, a process called *shotgunning*, the entire chromosome can be made to give up its secrets in relatively short order.

There are other more tangible benefits from this technology. *Recombinant DNA*—the name given to the battery of techniques by which the genes of several organisms are recombined in one—can be used to produce any kind of protein molecules for which we can find a gene, and in just about any desired quantity. Why would we want to produce protein molecules? The enzymes produced in a cell may be for the good of that cell, but they are not without other uses. Antibiotics, for instance, are chemicals produced by certain bacteria to eliminate competition from other bacteria, but they also have

The basic technique of recombining DNA molecules involves three separate operations: (1) In the first stage, the experimenter isolates and reconstructs the desired gene from the donor. To do this, the RNA that carries the message (mRNA) for the desired protein product is isolated. Double-stranded DNA is reconstructed from the mRNA. And, finally, the enzyme *terminal transferase* acts to extend the ends of the DNA strands with short sequences of identical bases (in this case, four guanines). These are going to be the bonds that hold the gene in place in its new molecular home.

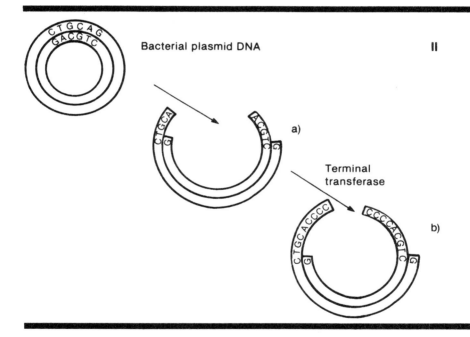

Bacterial plasmid DNA

II

a)

Terminal transferase

b)

(3) When you let these two get together, a bacterial plasmid containing the new gene is obtained. The plasmid can then be inserted into a bacterium, where it can be replicated. This will produce whatever protein the gene you chose produces—whether it's human insulin, growth hormone, or the ability to digest cellulose and produce alcohol.

SOURCE: Office of Technology Assessment.

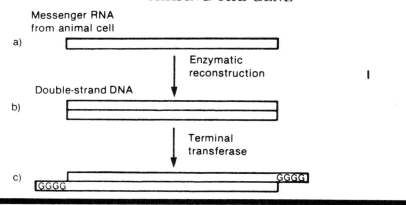

a) Messenger RNA
from animal cell

Enzymatic
reconstruction

I

b) Double-strand DNA

Terminal
transferase

c) GGGG   GGGG
GGGG

(2) A small piece of circular DNA called a *bacterial plasmid*
will serve as a vehicle for plugging the newly-isolated gene into
your bacterium. First, the circular plasmid is popped open by
using the appropriate *restriction enzyme.* The enzyme terminal transferase then extends the DNA strands of the broken
circle using identical bases. In this case, four cytosines are
used to pair off with the four guanines on the DNA strand
obtained in step 1. This is called *complementary base pairing.*

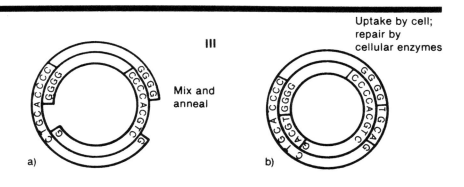

III

Mix and
anneal

Uptake by cell;
repair by
cellular enzymes

a)   b)

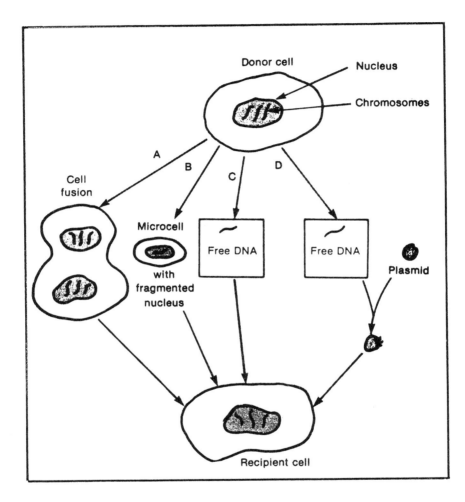

There are a number of ways that DNA can be transferred from one cell to another: (A) The two cells can be fused together completely; (B) A microcell with a fragmented nucleus can carry the DNA; (C) Free DNA can be injected by a virus or stumble into the cell with the aid of some calcium, or it can fuse to a cell after being coated with a naturally occurring membrane; (D) The free DNA can be joined to a plasmid and transferred as recombinant DNA.

You should note that there are a number of ways for DNA to get from one cell to another—and only one of them takes the help of a human.

wide medical applications—for killing bacteria. Beer and wine represent molecular by-products of certain bacterial metabolic processes. If the molecules produced by bacteria can be this useful, how much more useful might it be to insert the genes for other nonbacterial molecules into bacterial cells so that they can produce these as well?

That, in a nutshell, is the main thrust of recombinant DNA technology—the use of bacteria for manufacturing useful protein molecules, such as insulin, interferon, and growth hormones. And, because we have the opportunity to study these genes before and after splicing them into the bacterial hosts, it is even possible that we may alter their structures in some way to make them still more useful.

There is no question that this is a technology of inestimable value. But it has a dark side as well, and for a time in the 1970s it looked as though the age of genetic engineering would end as swiftly as it had begun.

DARK DAZE

Why did Paul Berg refuse to insert his recombinant plasmid into a bacterial cell?

In retrospect, it would seem almost an act of fate because of the complex sequence of events that this simple act unexpectedly set in motion and because it would not have happened at all if Berg, a professor at the Stanford University Medical Center, had not had a graduate student named Janet Mertz, who in 1971 attended a workshop on tumor viruses at Cold Spring Harbor Laboratory on Long Island, New York.

Cold Spring Harbor is one of the premier biological laboratories in the United States; it is directed by James Watson, the same Watson who, with Francis Crick, unraveled the secrets of the double helix in the 1950s. The tumor workshop that Mertz attended was taught by a young microbiologist named Robert Pollack, who had a particular interest in the ethics of science and scientists. The subject of the workshop was laboratory safety as it related to tumor virus research.

Pollack learned from Mertz of the SV40 experiment. While both Berg and Mertz saw the experiment as an elegant means of probing the genetics of viruses and bacteria, Pollack saw it as a perilous venture into an area where scientists had best tread cautiously, if at all.

Pollack's reasoning was this: The SV40 virus had been shown to cause tumors in certain animals and in laboratory cultures of human cells (though there was no evidence that SV40 or any other virus caused cancer in living human beings). How the virus induces tumors was not understood. One possibility was that the tumor virus, in inserting its genes into the chromosomes of animal cells, accidentally disrupts the normal genetic control processes of the cell, triggering a period of wild multiplication and growth.

The bacterium into which Berg intended to insert the SV40 was E. coli, which, as noted earlier, normally is found in the human intestines. Berg had chosen E. coli because, of all organisms known to man, the genetics of E. coli were the best understood, the bacterium having long been a favorite of geneticists, and because it was the organism with which, through many years of familiarity, biologists such as Berg felt most comfortable. But what, Pollack wondered, if a bacterium containing the SV40 DNA should somehow escape from its culture dish and infect a worker in Berg's lab? In the warm environment of the lab worker's gut, the SV40 DNA would have ample opportunity to work its carcinogenic wiles on human tissue. Who was to say that the combination of SV40 and E. coli DNA might not produce an organism more virulent than either? Furthermore, the hybrid E. coli would be quite capable of multiplying in huge numbers, and the hybrid DNA would be replicated each time the bacterium reproduced. If it should indeed become virulent, its virulence would be turned loose upon the world in quantity. Alternatively, the E. coli might pass its recombinant plasmids to other bacteria in the gut, which themselves might become capable of inducing cancer; the result could be a plague deadlier than any previously known.

Pollack found this scenario frightening, but when he called Berg at Stanford the senior biologist was unimpressed. Berg suggested that Pollack's fears were hypothetical and largely groundless and that they hardly outweighed the value of the experiment itself. But as Pollack pressed, Berg agreed to postpone the experiment until he had spoken to others about the dangers Pollack envisioned.

Berg did so, and he was taken aback by what he discovered. While no one could state unequivocally that Pollack's scenario actually could happen, no one was sure that it couldn't. The problem was chiefly lack of knowledge; recombinant DNA had taken biologists into uncharted waters, and no one knew just what dangers lurked there. Anything might happen. More information was required, but that information needed to be obtained cautiously.

Berg canceled his experiment. "I did not want to do an experiment that we might regret," he said later. "Once done it could not be undone. I was not absolutely certain that there was no risk."

Berg recontacted Pollack and offered to organize a pair of meetings to be held at the Asilomar Conference Center, a renovated chapel overlooking the ocean in Pacific Grove, California. There scientists involved with tumor research and genetic experimentation would be able to interact and to assess the hazards involved.

The first of these meetings, attended by one hundred tumor specialists from around the world, was held early in 1973. The results were inconclusive. Toward the end, Berg made a plea for "a study, most logically sponsored by or even carried out within the National Institutes of Health [NIH], to determine if there is increased risk of cancer stemming from current and projected researches with biologic oncogenic [tumor-causing] agents."

The call went unheeded. In fact, in retrospect, perhaps the most striking thing about the first Asilomar Conference is how little did come out of it. There was no press coverage; only poorly transcribed accounts of the panel discussions survive.

Yet, in 1973, the newborn science of genetic engineering stood on the threshold of one of the greatest controversies the biological sciences have yet known, and the very name Asilomar was to become a byword in a heated debate over scientific ethics and responsibility.

The engineers of life were about to be brought to account.

EXPERIMENTS AND ARGUMENTS

Six months after the first Asilomar Conference, the events that would eventually come to a head at Asilomar II were set in motion.

The Gordon Research Conference began in 1931 as a series of annual gatherings where scientists could discuss their research in a relaxed, informal atmosphere. By 1973, the conference had been renamed the Gordon Research Conference on Nucleic Acids; attendance was by invitation only.

Attending the 1973 conference was Herbert Boyer, who, along with Stanley Cohen, was then involved in the early stages of the first experiments with recombinant DNA. One cochairperson of the conference was Dr. Maxine Singer, a chemist with the National Institutes of Health. Singer was a personal friend of Paul Berg and one of the scientists with whom Berg had discussed the possible consequences of the SV40 experiments.

Thus, when Boyer revealed that he and Cohen had developed a method by which diverse genetic materials could be brought together in a single organism, Singer recognized the technique as a refinement of that used by Berg—and one that presented the same possibilities of hazardous recombinations of genes; such recombinations, in fact, would be made easier than ever with Boyer's methodology. If genes from higher organisms now could be spliced routinely into bacteria, Singer reasoned, it was important that those involved, or potentially involved, with this research understood the kinds of dangers it might entail. Singer discussed this with cochairperson Dieter Soll, a biologist from Yale. They agreed that they would bring the issue before the conference.

Singer hastily drafted a statement, which she read the next morning. Unfortunately, by that time roughly a third of the group had gone home. Those who remained voted to send a letter of concern, to be signed by Singer and Soll, to the National Academy of Sciences in Washington. The letter was for publication in the magazine *Science*, published by the academy. (Specifically, seventy-eight of the remaining ninety-five scientists voted to send the letter, though only forty-eight voted for its publication.)

The letter appeared in the September 1973 issue of *Science*. In fairly terse scientific phraseology, it outlined the techniques that Boyer had discussed at the Gordon Conference and the possible hazards presented by their use. The letter would not be understood by the average reader—it contained phrases such as "overlapping sequence homologies" and "covalent linkage of molecules." It suggested that:

> . . . hybrid plasmids in viruses, with biological activity of unpredictable nature, may eventually be created. . . . Certain such hybrid molecules may prove hazardous to laboratory workers and to the public. Although no such hazard has yet been established, prudence suggests that the potential hazard be seriously considered.

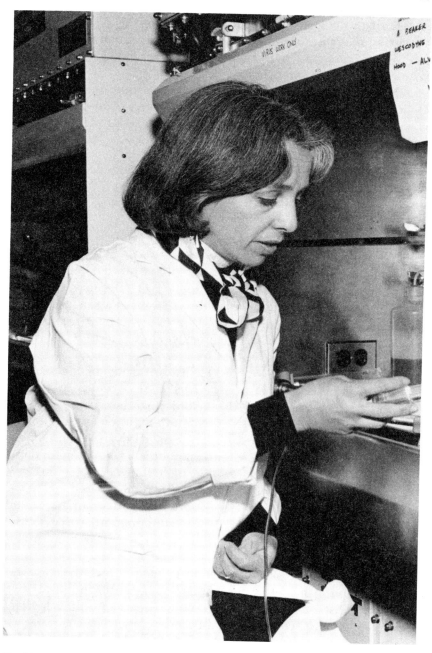

Dr. Maxine Singer was the first to question *formally* what the potential hazards of this new technology were. Her questions led to Asilomar II. *(courtesy National Institutes of Health)*

Perhaps the key sentence in the letter was this: "The conferees suggested that the Academies establish a study committee to consider the problem and to recommend specific actions or guidelines, should that seem appropriate."

The National Academy responded by setting up an informal committee to investigate the possible hazards of recombinant DNA research. To chair the committee, the National Academy's president, Philip Handler, turned to the man who had started the sequence of events: Paul Berg.

BIOHAZARDS

The committee that Berg gathered about himself consisted of eight prominent biologists: Berg himself, David Baltimore, Herman Lewis, Daniel Nathans, Richard Roblin, James Watson, Sherman Weissman, and Norton Zinder. After a brief preliminary meeting in Washington, D.C., they met in Cambridge, Massachusetts, in April 1974.

The Berg committee worked to firm up plans for the second Asilomar Conference, which they envisioned as a chance for researchers from around the world to come together and assess the risks—the so-called biohazards—surrounding certain kinds of genetic research. Safety guidelines could be drawn up at the conference and an agenda devised for further exploration of the potential dangers presented by recombinant DNA techniques.

What the Berg committee will best be remembered for, however, is a single act: the calling of a moratorium on several types of research. The committee members feared, apparently, that in the year before the conference was held some few researchers might engage in an orgy of dangerous gene-splicing experiments—getting them in, as it were, before the ax fell at Asilomar. Thus, the committee voted to publish a letter in several major scientific publications, asking scientists to refrain from such research until tentative guidelines, at least, could be established.

The letter, signed by Paul Berg et al. (and which came to be known, within the scientific community, as the Bergetal letter), was published in July 1974. Whereas the Singer-Soll letter was terse and largely incomprehensible to the uninitiated, the Bergetal letter was relatively clear and expansive. It read:

> Several groups of scientists are now planning to use this [gene splicing] technology to create recombinant DNA's from a variety of other viral, animal and bacterial sources. Although such experiments are likely to implement the solution of important theoretical and practical biological problems, they would also result in the creation of novel types of infectious DNA elements whose biological properties cannot be completely predicted in advance. . . . There is serious concern that some of these artificial molecules could prove biologically hazardous.

Whereas Singer-Soll had merely called for the creation of a committee to assess the problem, the committee itself recommended more drastic action:

> First, and most important . . . until the potential hazards of such recombinant DNA molecules have been better evaluated or more adequate methods are developed for preventing their spread, [we ask that] scientists throughout the world join with the members of this committee in voluntarily deferring [certain] types of experiments.

The types of experiments to be deferred were those in which genes for antibiotic resistance were placed in bacteria (or plasmids) that didn't already have those genes; and those in which genes from tumor-causing viruses were placed in bacteria or (plasmids).

The Bergetal committee had expected a storm of protest from within the scientific community; the call for a moratorium was drastic and unprecedented. Recombinant DNA research was a burgeoning field and biologists would be reluctant to drop their research.

What the Bergetal committee had not expected was a storm outside of the scientific community, but that was what they got. And it began to break the day the letter saw print.

The relationship between science and society always has been a tenuous one, ill drawn and ill understood, always changing. As the role of the scientist has become better defined—and as the scientist himself has become more important to society as a whole—so has the scientist taken on an increasing burden of responsibility; he has become answerable for his acts. This is a burden that some scientists have accepted only grudgingly.

As recently as the past century, the scientist had no particular role in society. As June Goodfield notes in her book *Playing God: Genetic Engineering and the Manipulation of Life:*

> There was a time when the question, "Who is a scientist?" was meaningless, not only because science was not a recognized professional activity, but because the very word itself had not been coined. During the Seventeenth and Eighteenth Centuries—indeed, during the first few years of the Nineteenth Century—those natural philosophers, the people whose activity we recognize as scientific, were usually clergymen, like Stephen Hales, or tax collectors, like Antoine Lavoisier. It was an amateur pursuit, open to anyone who could support himself, and engaged in with varying degrees of seriousness or light-heartedness.*

Even at the turn of the twentieth century, the entity that we call science bore little resemblance to that entity today: There were no huge budgets for scientific research, no sprawling university laboratories, no thousands upon thousands of scientific journals. "Big science," as the modern research establishment is sometimes pejoratively called, did not yet exist.

But the twentieth century has been the century of technology, the century of invention, and whereas two hundred years ago the scientist as such did not even exist, today he is preeminent. The modern researcher—in biology, in physics—wields a powerful sword of influence, but it is a two-edged sword. With this power has come a loss of innocence. When science was a pursuit for light-hearted amateurs, the scientist was at least free to investigate what he wished, pursuing those leads he thought valuable. As the scientist's influence waxed, this autonomy has waned. One reason is, simply, money: Physicists require tens of millions for some experimental machines, and biologists the odd million for a proper lab. So, much modern research requires that the researcher justify himself periodically to whoever pays the bills, be it government or private industry. But increasingly it has become incumbent on the scientist—or at least some scientists—to justify his work to the public, to prove that it is in the general interest.

Influence and loss of autonomy: If any single event is symbolic of this dichotomy, it is the dropping of the atomic bomb on

*© Goodfield, June. *Playing God: Genetic Engineering and the Manipulation of Life*. New York: Random House, Inc. 1977.

Hiroshima. In a single searing flash of gamma rays, modern science proved that it was capable of altering the universe, or at least our small portion of it, but in a particularly ghastly way.

In the wake of Hiroshima, science budgets soared; huge laboratories were built; lavish funds were apportioned. But all at once the question on the lips of concerned citizens was not "Can science change the world?" but rather "Do we want science to change the world?"

Little of this new consciousness reflected on biology. Unlike physicists, biologists could not split the atom. There was nothing to fear from Petri dishes and bacterial cultures. Oh, there was an occasional accident, when some virus managed to escape from a test tube and infect a lab worker, but that was hardly the biologist's fault; the virus had existed before it was placed in a test tube, and the virus was quite capable of virulence on its own. Atoms, on the other hand, did not split without encouragement.

Genetic engineering was something else again. All at once, biologists had been given a tool as powerful in its way as the atom smasher, a tool with which they could tamper with the primal stuff of life and reassemble living molecules in dangerously unpredictable ways.

For physicists, the road to Hiroshima had been paved with good intentions. The atomic bomb was developed in a time of war, in the fear that Hitler, with his vast reservoir of brilliant physicists (Germany having been the source of so much revolutionary twentieth-century physics), was ready to tap into the heart of the atom. It was dropped in the heated urgency of the bloody end game of the war with Japan, when it was widely believed that any extension of the war would mean the lives of thousands of Allied soldiers. It is easy in retrospect to question the necessity of Hiroshima, to ask why no one stopped the process when it could have been stopped; but the historical momentum seems to have been overwhelming.

Yet, confronted with their own potential Hiroshima, the molecular biologists did something perhaps unique in the history of science: They pulled back from the brink and began to examine the nature of the precipice. The Bergetal letter was received with bitterness in some quarters, but by and large it was obeyed. And respected.

What surprised the biological community—though, on reflection, it seems almost inevitable—was that the general public met the moratorium not with applause but with distrust.

ASILOMAR II

The first Asilomar Conference had gone largely unnoticed by the general public; so had the Gordon Conference letter. As far as most biologists were concerned, the recombinant-DNA debate was an internal affair of the scientific community, and no one else's business.

This, too, was how the Bergetal committee viewed the upcoming Asilomar Conference—as a chance for scientists to join together and discuss matters of interest solely to other scientists, to address a set of purely technical issues that could then be clarified and resolved in a largely technical manner.

It didn't happen that way. On the day that the moratorium was announced, newspapers across the country trumpeted the news, and more than just scientists were listening. "World Health Peril Feared" blared the headline in the *Philadelphia Inquirer*. "Risk to Man Seen in Creating New Bacteria" shouted the *Times* of London.

Whatever hopes that biologists might have entertained for keeping the recombinant-DNA question within the scientific community were dashed. The very idea of creating new life forms struck a resounding note (and perhaps a nerve) with both press and public. The popular analogy for recombinant-DNA research became not Hiroshima but *The Andromeda Strain*, Michael Crichton's best-selling novel about a rapidly mutating microorganism brought to earth on a returning space probe with disastrous consequences. Gene splicing had about it a distinct aura of science fiction—and, to a public weaned on *The Outer Limits* and *The Blob*, most science-fiction scenarios were dark ones.

The International Conference on Recombinant DNA Molecules—Asilomar II—was held from February 24 to 27, 1975, at the same conference site in Pacific Grove as the first conference. One hundred forty scientists attended, ninety of them American, fifty of them from countries as far flung as the Soviet Union and Japan, Australia and Belgium. The press was there as well; after the publicity accorded the moratorium it would have been difficult to keep them away. One article covering the conference, published in *Rolling Stone*, pithily characterized the proceedings in its subtitle: "140 Scientists Ask: Now That We Can Rewrite the Genetic Code, What Are We Going to Say?"

Ostensibly, the conference was to give researchers an opportunity to draw up guidelines for dealing with potentially hazardous materials. But much of the debate revolved around whether such hazards in fact existed. James Watson, one of the original members of the Bergetal committee, startled conference goers by declaring that the moratorium should end immediately because there was no significant hazard in the research (a position that Watson came to hold shortly after the Bergetal letter was published and apparently maintains to this day). "The dangers involved," he said, referring to gene-splicing technology, "are probably no more dangerous than working in a hospital."

Judging from transcripts of the Asilomar debates, Watson's was a minority view. Most of those attending agreed that there might be hazards involved in the work but that they would be difficult to quantify. For instance, what was the actual likelihood that a laboratory strain of *E. coli* would colonize the intestines of a laboratory worker? And could genes transplanted into that *E. coli* actually render this normally harmless bacterium virulent?

These were important questions, and the answers were not

James Watson (left) argued for the end of the experimental moratorium. While at Asilomar he and Sydney Brenner (right) discussed the debate. *(courtesy National Academy of Sciences)*

easily come by. Until Asilomar, it had been widely assumed that the specially bred strains of *E. coli* used in the laboratory were no longer capable of living "in the wild"—that is, in the ordinary habitat of nonlaboratory strains of *E. coli* (the so-called wild-type strains). Furthermore, new strains of *E. coli* even feebler than those currently in use could easily be bred—strains, for example, that would require a constant supply of chemicals in order to survive, chemicals not ordinarily found in the human body; strains that could not survive at bodily temperatures; in short, strains that would die the moment they left the delicately controlled environment of the laboratory.

Even if a supposedly enfeebled bacterium did manage to escape the lab and survive in the wild, no one knew if the recombined genes inserted into that bacterium could have any effect on a human being. Would the broken and shotgunned genes of a cancer-inducing virus still be capable of entering a human cell, much less of causing that cell to become malignant? Could smallpox genes in *E. coli* still cause smallpox? Would the genes from the organism that creates botulism toxin still be poisonous inside a bacterium?

There was no evidence that they could; on the other hand, there was no evidence that they could not. Once again, no one knew. At Asilomar, it was agreed that experiments should be performed to assess the biochemical activity of these and other recombinant creations, provided such experiments were carried out with an excess of caution. Until the results from such experiments could be evaluated, certain potentially dangerous experiments would either be deferred or be performed under highly restrictive conditions.

One particular fear expressed repeatedly at the conference was that, in the words of Bergetal committee member Daniel Nathans, "molecular biologists were not trained to handle organisms safely." Until the 1970s, viruses and bacteria had been largely the province of the microbiologists, who were trained in proper laboratory practice practically from the first time they touched a test tube. As Nicholas Wade, a *Science* magazine correspondent at Asilomar, noted, the microbiologists at the conference were identifiable "by their habit of using their elbows to shut off washroom faucets." To the microbiologists, the "molecular people" were sloppy and badly trained; they could not be trusted to handle virulent organisms without extensive retraining in sanitary techniques.

Two years after the conference, Richard Novick, a biologist with the Public Health Research Institute, made this point more strongly: "Most of those for whom the guidelines [rough-drafted at Asilomar] are intended," he wrote, "are not now trained in micro-biological safety techniques. In fact, most of them have chosen to work with *E. coli* and other relatively innocuous microorganisms precisely because with such organisms one didn't have to worry about contamination, dissemination and infection. . . . The average biochemist-molecular biologist is not likely to have the training and tradition of safety awareness either to perceive the various deficiencies or to exercise optimum discretion." It was generally agreed that something would have to be done about this, though no one was quite sure what.

From the viewpoint of the scientists present, perhaps the most stunning moment of the entire Asilomar Conference came on the final evening, when three lawyers presented, in turn, their views of the legal implications of the research. It was here, for the first time, that the potentially explosive issue of scientists' rights was broached. How much right did scientists have to perform experiments, no matter how valuable in terms of knowledge gained, if those experiments were potentially dangerous to the public welfare? None at all, asserted Roger Dworkin, a professor of international law at the University of Indiana Law School.

According to Dworkin, whatever freedom scientists have traditionally enjoyed in pursuing their interests, in regulating those interests, and in deciding what is safe and what is not safe in their experiments, has been bestowed on them by the public, and they enjoyed this freedom only on public sufferance. If the public should decide to revoke that freedom, that is its privilege.

And what if the public should abuse that privilege, wield it unwisely? "It is the right of the public to act through the legislature to make erroneous decisions," said Dworkin.

Those words were to echo loudly in the months and years to follow. If scientists did not police their own research—in an obvious and public way—then the people would do it for them.

And that was enough to strike fear into the heart of even the most conscientious scientist.

RECOMBINANT RULES

Ultimately, the Asilomar Conference produced a document which suggested a hierarchy of safety procedures to be used in laboratories working with recombinant organisms. The document was drawn up by committee and debated somewhat hastily by those in attendance, some of whom grumbled that it had been railroaded through under less than ideal conditions.

The document suggested that there be two types of confinement procedures—that is, two types of procedures designed to confine the activity of recombinant organisms to the laboratory: physical and biological. Physical confinement would involve the actual restraint of the organism: retaining it behind glass and steel partitions, specially designed ventilation systems and air locks; requiring lab workers to shower (and perhaps even submit to a physical examination) before entering or leaving a laboratory where recombinant DNA experiments were in progress. Such confinement procedures already existed, mostly for use in laboratories dealing with germ warfare.

Biological confinement, on the other hand, involved the development and use of deliberately enfeebled microorganisms that stood an infinitesimally small chance of surviving outside the laboratory. Many of those at the conference believed that biological confinement would be much more effective than the physical kind.

Each type of confinement was to operate on several different levels, the particular level varying according to the degree of risk believed to be associated with the type of experiments being performed. Experiments deemed particularly risky included those involving genes from harmful viruses, or toxin-producing organisms, or shotgun experiments involving organisms biologically close to humans (on the assumption that there might be dangerous viral organisms lying dormant in these genes). Experiments involving highly pathogenic organisms (based on existing classifications for pathogenicity) were deferred altogether, as were all large-scale experiments (which the conference defined as those involving more than 10 liters of bacterial culture).

A few months before Asilomar, in October 1974, the National Institutes of Health had formed the awkwardly named Recombinant DNA Molecule Program Advisory Committee (later shortened to the Recombinant DNA Advisory Committee, better known as RAC). Using the Asilomar document as a springboard, the RAC began drawing up a set of official guidelines for recombinant DNA research. In February 1976, a year after Asilomar, the guidelines appeared in draft form, for public debate.

And debate there was.

THE GREAT DEBATE

The public debate over recombinant DNA research began before the RAC guidelines were released and continued long after those guidelines appeared in final (though eventually to be revised) form. The debate turned largely on two issues: Was there significant danger in genetic experimentation? (This question was not satisfactorily answered at Asilomar.) Were scientists (who could hardly be considered disinterested) the right ones to decide? The debate was hardly cooled by the fact that the Recombinant DNA Advisory Committee was made up almost entirely of the aforementioned nondisinterested scientists, a situation that was later remedied, as nonscientists were admitted to the committee.

The guidelines made specific the confinement levels suggested by the Asilomar document. As devised by the RAC, the categories for physical confinement were these:

P1—requiring only ordinary microbiological laboratory practices.

P2—same as above, but with restricted access to the laboratory (which means, essentially, hanging a KEEP OUT sign on the door).

P3—same as above, but with the laboratory air pressure kept lower than that of its surroundings, so that airborne microorganisms will blow into the laboratory rather than out of it.

P4—high containment, similar to that used in biological warfare experiments, requiring laboratory workers to change clothes and shower when entering or leaving, an air lock between the lab and outside, etc.

The levels of biological confinement became:

EK1—using standard laboratory *E. coli* (a strain known as *E. Coli* k-12, developed some decades earlier).

EK2—using specially designed strains of *E. coli*, believed to have one chance in 100 million of surviving outside the laboratory.

EK3—using specially designed strains of *E. coli*, demonstrated (through extensive experimentation) to have one chance in 100 million of surviving outside the laboratory.

(Interestingly, no strain of *E. coli* qualifying for EK3 confinement ever has been developed, though at the time the guidelines were developed it was believed that one soon would be.)

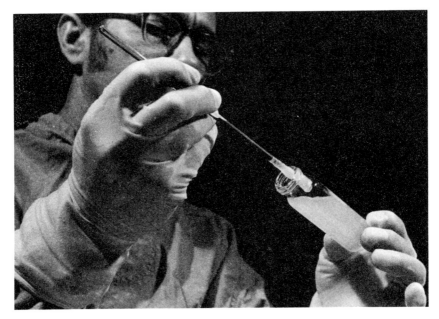

DNA of a bacteria culture being prepared under P3 conditions.
A glass rod is used to spin the DNA into a cohesive mass after
it has been separated from other materials such as protein.
(courtesy National Institutes of Health)

The guidelines included a detailed list of what kind of experiments qualified for what confinement levels. These were criticized, in the early drafts at least, as being somewhat lenient; the members of the RAC countered that if they had been much more stringent, many biologists might have chosen to ignore them entirely. The National Institutes of Health were the enforcing body behind the guidelines, and their only means of enforcement was withdrawal of funds for noncompliance, which meant that the guidelines applied only to those researchers receiving government funds. Though this covered virtually all university research, genetic experimentation by private industry was effectively unrestricted, although voluntary compliance seemed to be the rule. (In a later version of the guidelines, specific requirements for voluntary compliance were inserted and largely adhered to.)

Whether the guidelines were adequate, too lenient, or too stringent still depended on whether the dangers associated with recombinant DNA were real or imaginary.

Robert Sinsheimer, a biologist at the California Institute of

Technology in Pasadena, was one of the most impassioned critics of genetic engineering. At one time he had been involved with the work that eventually led to gene splicing, but with the actual arrival of this technology, he had come to have second thoughts, based not so much on any specific technical problems that he foresaw in the experimentation, but on this general principle: Man should cross evolutionary barriers only with the greatest trepidation.

This is no small consideration. For the most part, gene splicing involves the insertion of eukaryotic genes into prokaryotic cells; the difference between these two types of organism is a fundamental biological barrier (or so it seemed in the early 1970s), a barrier that must exist, Sinsheimer reasoned, for valid evolutionary reasons.

Even the insertion of two genes from different species into a single organism would represent the breaching of an evolutionary wall: The very definition of species, in biology, hangs on the ability of individuals to mate—to mix genes and produce offspring. If two individuals can mate (and produce an offspring that is not sterile), then they are members of the same species. If they cannot, they are not. Yet, with recombinant DNA technology, the mingling of genes between species is not only possible but even common; no evolutionary barriers stand in the way of gene splicing. But perhaps they should. . . .

Sinsheimer wrote in the British journal *New Scientist:*

> Nature has, by often complex means, carefully prevented genetic interactions between species. Genes, old and new, can only interact within a particular species.
>
> With recombinant DNA research we can now transform the evolutionary tree into a network. We can merge genes of most diverse origins—from plant or insect, from fungus or man, as we wish. Most such combinations will, of course, be sheer nonsense, non-viable and innocuous. A few will, by careful design, be valuable; if not, there would be little point to the whole enterprise. A few, however, may by design or inadvertence be deadly, in any of many ways.

Sinsheimer goes on to point out that nature has discarded any number of species in the past as unsuccessful experiments. Presumably, among those discarded species, may be several that developed the deadly ability to commingle genes across evolutionary barriers. Since evolution functions by ruthlessly pruning the evolutionary tree of nonviable species, it can be assumed that if certain abilities do not exist in nature, then probably they have been tried and discarded. If evolutionary barriers cannot be crossed naturally, then they should not be crossed. The penalty for trying is being selected out of the gene pool.

But is it true that these supposed evolutionary barriers have never been broken? Sinsheimer's argument turns on this assumption. Yet in the wake of Asilomar—during the many experiments that followed the conference—much evidence to the contrary began to emerge. There is, for instance, considerable evidence that the *E. coli* in the human intestines is frequently receiving infusions of human genes from the intestines themselves. Microorganisms in the soil may be subject to a similar process, by which decaying plant and vegetable matter ooze their genes past the supposed evolutionary barriers and into the chromosomes of bacteria. It even can be argued that such interspecies genetic transfer is an important part of the evolutionary process, a means of spreading sequences of genes far afield to see if valuable new combinations can be created, just as bacteria exchange plasmids to spread valuable combinations among themselves. In the words of bacteriologist Bernard D. Davis, of Harvard Medical School:

Bacteria in nature have long been exposed to DNA from lysed [partially dissolved] mammalian cells—for example, in the gut and in decomposing corpses. *Escherichia coli* can take up DNA after damage to the cell envelope, and one would expect . . . such damage occasionally . . . Over the past 1,000,000 years human-bacterial hybrids are exceedingly likely to have already appeared and been tested in the crucible of natural selection. If so, experimental DNA recombination will not be yielding a totally novel class of organisms.

James Watson chimed his agreement in the pages of *The Bulletin of the Atomic Scientists:*

My thinking at Asilomar was strongly influenced by the relative ease with which DNA molecules can pass into cells and be incorporated in a genetically active form into the chromosomes of the new hosts. When this phenomenon was first discovered in 1943 [by Oswald Avery], it was thought to be restricted to the pneumococci bacteria. But over the years we have learned of more and more experimental cases where genetically active DNA has been transferred to a variety of different bacteria as well as to the cells of many higher organisms. That being the case, it seems probable that such transfers will also occasionally occur normally *in vivo.*

The bacteria in our body, for example, must constantly be exposed to DNA released from either sick human cells or from free

DNA present in our food. If a person is on antacid pills, there is a good probability that his intestinal bacteria would be exposed to and take up a tiny fraction of the DNA present in, say, raw oysters or beefsteak tartar. And even without the antacid, the very rare DNA molecule must pass through the digestive tract to reach not only our intestinal bacteria but, also, the cells lining the gut.

How infrequent such events are is not yet known. But we should assume that on an absolute scale the transfer of genetically active DNA is a very common event, and that it generally occurs with no harmful consequence to host cells already well adapted to their own genetic messages. As a result, I see no reason to be apprehensive about any experiment in which we transfer genes of higher cells to bacteria.

Nature, then, is constantly performing its own recombinant DNA experiments—a powerful argument against Sinsheimer's disaster scenarios.

The critics of gene splicing offered other hypothetical disasters, however, some of them not so easily answered: the cellulase scenario, for instance.

In a P4 containment lab, all the work takes place inside the system of enclosed gas-tight cabinets you see here. *(courtesy National Institutes of Health)*

There is an enzyme called cellulase which, as its name implies, breaks down cellulose molecules. It is found in certain bacteria, and it is these bacteria, living in symbiosis with animals known as ruminants (the cow is one), that allow these animals to derive nourishment from cellulose fibers, such as those in grass. Human beings have no such symbiotes and lack the ability to synthesize the enzyme; therefore we cannot digest cellulose.

Under certain circumstances, it might be a considerable boon to be able to digest such fibers—if other foods were in short supply, say. In a time of famine it might occur to some benevolent professor of biology to splice into some harmless bacterium (such as *E. coli*) the gene for cellulase and then insert that bacterium into a human being; an individual equipped with the cellulase bacterium then would be able to digest cellulose fiber much as a cow does. Deprived of other nourishment, the recipient could graze contentedly in his backyard while all about him or her starved.

But would the recipient really be better off? The human digestive system requires a certain amount of undigested fiber if it is to function properly; lack of this fiber could lead to cancer of the colon and rectum or even terminal constipation. With the cellulase enzyme functioning in one's gut, little fiber could get through without being digested; the result would be disastrous.

Of course, knowing this, no one would be silly enough to introduce such a cellulase-producing organism into the human body.

But who is to say that other uses for recombinant organisms might not have equally disastrous consequences, as superficially appealing as they might appear at first blush? We just don't know; there is too much that we do not understand. A potentially attractive genetic experiment might have cataclysmic ecological repercussions, but we would not know it until it was too late. Our ignorance simply is too vast.

And what about the autoimmune disease scenario? When the body is invaded by foreign particles—bacteria, viruses, whatever— it strikes back with tiny weapons called antibodies, molecular structures produced by the blood cells to track down invading particles. Each of the thousands of antibodies in your body is designed to "recognize" the distinctive physical pattern of a certain kind of intruder; it does this by virtue of its shape, as a lock "recognizes" a key. When an antibody finds one of the intruders for which it is

Scientists can move and control all the materials used in P4 experiments using shoulder-length heavy gloves attached to openings in the cabinets without ever having direct contact with anything inside the cabinets. Under current NIH guidelines, P4 containment no longer is required, though the facility still is available for conducting risk-assessment experiments and for research involving highly pathogenic organisms. *(courtesy National Institutes of Health)*

specific, it attaches itself to the particle, rendering it harmless or summoning aid from other soldiers in the body's army of immunity.

If the body lacks an antibody specific for the invading particle, it makes one; first, however, the intruder has to be identified as foreign so that the antibodies will not accidentally attack part of the body itself.

Sometimes this system goes awry, and an antibody forms that is specific for the body's own tissues. In arthritis, for instance, antibodies are produced against the connective tissues in the joints, and they attack those tissues, creating an inflammation. Such a condition is called an autoimmune disease. It occurs when the immune system turns on the body it is supposed to be protecting.

It is possible that a recombinant organism containing human genes, if introduced into the human body, would be recognized as foreign matter and therefore would stimulate production of antibodies against the proteins coded for by the recombinant genes. Because these proteins would be identical to those produced by the body itself, an autoimmune condition could result. The antibodies thus produced might attack the body's own proteins.

Is this possible? Is it likely? No one can say that it isn't.

What makes these scenarios—and others like them—all the more terrifying is that recombinant organisms, unlike ordinary poisons and pollutants, are self-replicating; they make copies of themselves. While the particulate matter that makes up ordinary pollution might eventually flush itself out of the biosphere, a living organism—a living pollutant—might prove difficult, if not impossible, to eradicate. A bacterium that secreted human insulin, for instance, might escape from the laboratory and produce an epidemic of hyperinsulinism. Who can say what might become deadly if reproduced in great enough numbers?

It also can be argued, however, that a recombinant organism has far less chance of surviving in the wild than does a wild-type organism, and not just because it is of a weaker strain. As far as a bacterium such as *E. coli* is concerned, the extra genes introduced by biologists are just so much cumbersome excess baggage; a bacterium carrying these unnecessary genes would function less efficiently than a wild-type bacterium and would quickly lose out in any evolutionary competition. In short, recombinant bacteria are unlikely to survive and multiply in the wild simply because they are recombinant—and if you add to that the fact that they were engineered from enfeebled strains to begin with, the likelihood of some dangerous bacterium proliferating outside the laboratory seems vanishingly small.

Within a year or so of the Asilomar Conference it had become obvious to many scientists that disaster scenarios painted by biologists before the moratorium had little if any basis in fact. Though no one could prove that there was no danger involved in such experimentation, it soon became evident that these dangers must have at least a low level of probability. K-12 *E. coli* seemed unable to survive outside the laboratory. Human (and other eukaryotic) genes frequently mixed with prokaryotic genes in a number of natural situations. Most significantly, no one seemed to be able to create a virulent strain of *E. coli*—not even deliberately.

Wrote David Baltimore, a member of the original Bergetal committee and of the RAC, in regard to the continuing restriction of recombinant DNA work:

> Since 1976 neither experimental evidence nor solid theoretical arguments have been advanced to support the position that recombinant DNA research poses any danger to human health or to the integrity of the natural environment. We doubt that the beneficial side effects of continued regulation justify the expenditure of time and money required to maintain a regulatory apparatus that has been developed to protect society from hazards that appear to be nonexistent.

Pathogenic genes consistently proved harmless, or relatively so, outside of their natural environments. This was the irony that made Asilomar, in retrospect, seem something of an overreaction: Recombinant DNA technology, applied to the study of pathogenic organisms, apparently was safer than the more traditional methods for studying these organisms. By removing the genes from their original hosts and shattering them into fragments, the pathogenic organisms were rendered less pathogenic, and the *E. coli* in which the genes were spliced remained harmless.

"Experience of the last six years," according to Paul Berg, "has convinced me that our earlier concerns are no longer warranted. I now believe there is more to fear from the intrusions of government in scientific research than from recombinant DNA experiments themselves."

Yet the public felt otherwise, and perhaps with some reason. Even as scientists were prepared to return to business as usual and to shuck aside Asilomar as a resolved issue, public resistance to recombinant DNA was just beginning to peak.

In an era of post-Watergate cynicism, the average concerned citizen saw the moratorium not as an act of civic responsibility but as a confession of dreadful guilt on the part of the scientific community; Asilomar was an attempt to cover up after the moratorium, to whitewash it. The more that biologists tried to explain that it was just the opposite, that the dangers under study at Asilomar had been purely conjectural, the more hollowly their words seemed to ring in the public ear.

"My phones didn't quit," says Maxine Singer, cochairperson of the 1973 Gordon Conference and cosigner of the original letter to the National Academy of Sciences. "I would patiently explain molecular biology research and explain that the scary, doomsday scenarios were pure fiction and why. It was frustrating and impossible. It was an absolutely astonishing nightmare."

Part of the problem stemmed from a public misperception that the moratorium had covered *all* recombinant DNA research, not just a few types of experiments. "There was and still is a misconception that we advocated a ban on recombinant DNA research," wrote Paul Berg in 1977. "It was never true."

Stanley Cohen, writing in *Science* magazine in that same year, sounded a similar note: "Contrary to what was believed by many observers, our concerns pertained to a few very specific types of experiments that could be carried out with the new techniques, not to the techniques themselves."

In effect, biologists felt as though they had opened a Pandora's box that could not be closed; the public was frightened and angry and, given the public's distrust of the scientists' pronouncements, it could not be calmed with the voice of gentle reason. "The vision of the hysterics," wrote James Watson in the *New Republic*, "has so peopled biological laboratories with monsters and super bugs that I often feel the discussion has descended to the realm of a surrealistic nightmare from which we will most surely soon awaken."

Perhaps. But there was also a strong—and probably accurate—public perception that the average biologist would much rather sweep the whole issue under the carpet than put it out in the open, where it could be discussed intelligently.

By mid-1976 there were hearings under way in several states on the efficacy of establishing laws binding on DNA research (adding teeth, in effect, to the NIH guidelines), and a drive was afoot, spearheaded by Senator Edward Kennedy of Massachusetts, to produce national legislation on the subject. The proposal for national

It usually took two or more people to perform experiments because manipulations or movements inside the "cabinet line" were either too awkward or simply impossible. *(courtesy National Institutes of Health)*

legislation eventually came to nothing, but several local ordinances—in Cambridge and Amherst, Massachusetts; in Berkeley, California; in Maryland; in New York State and elsewhere—were passed.

In the end, the debate concerning recombinant DNA was not so much resolved as forgotten. With no concrete evidence that the consequences of such research would be disastrous, the voices in favor of continuing or strengthening the restrictions died away to a low murmur (though the murmuring is far from silenced).

By 1980, the scientific community had begun to chafe impatiently at the restrictions placed upon them by the NIH guidelines. Wrote Norton Zinder, another member of the original Bergetal committee:

> The guidelines have outlived any usefulness they may have had. Recombinant DNA research has proven itself to be as safe as any other biological research. The evidence for this lies in myriads of successful experiments and the failures of many theoretical arguments, no matter how rigorously constructed to prove otherwise.

As though in response, the RAC voted in 1981 to remove virtually all restrictions from gene-splicing research and to make the remaining restrictions voluntary.

The debate seems to have left a bitter taste in the mouths of some scientists, especially James Watson, who in time came to feel that Asilomar may well have been a mistake. In Watson's words: "What started out as an attempt of the scientific community to appear responsible took on increasingly the aspect of a black comedy."

It's a pity that he feels that way. For once, the scientific community acted as scrupulously as possible in the face of conjectural hazards, and most of those who acted that way now seem to regret it. For some scientists (and we can hope that they are few), the lesson of Asilomar seems to be that ethical problems—especially those that involve the public—are best ignored.

David Baltimore wrote in 1980:

> Many times over the last five years I have asked myself whether the Berg letter was a good idea or not. When I'm feeling selfish, my answer is that it was a poor idea. It has caused me and many of my colleagues to spend endless hours trying to convince others that the potential for hazard is nowhere near as great as they think. At times, when the pressures of the controversy lessen and I can be a bit more reflective, I think [it] represented a reasonable response. . . .

Baltimore's ambivalence about Asilomar would seem to reflect the ambivalence of the scientific community as a whole: Is it better to ignore the welfare of society in order that you may do your work unmolested? Or is it better to acknowledge a larger responsibility at the risk of being unable to work at all?

The controversy over recombinant DNA seems to have grown more out of the mutual distrust between scientists and nonscientists than it did out of any genuine and substantial threat. In their fear of public overreaction, certain members of the scientific community apparently believe that they may be better off if the public is not informed of conjectured dangers in their work. And, in their fear of scientists not informing them of conjectured dangers, the public sometimes overreacts. In so doing, each reinforces the other's worst fears.

This may be the moral lesson of Asilomar: If the public and the scientists cannot forge a chain of mutual trust, ultimately both will suffer. And the only way to forge such a chain is by more actions such as Asilomar, with the scientists showing their willingness to discuss important issues with the public, and with the public showing its willingness to listen.

GROWING YOUR OWN . . .
AND HOW IT WORKS

Out of the chaotic birth of gene splicing has risen, with surprising speed, both an industry and a body of knowledge. Perhaps this more than anything else is what has silenced the voice of controversy: There is money in genes. And a great deal more.

To appreciate how much more, let's pause for a moment to examine the techniques of recombinant DNA in somewhat greater detail. How would we go about creating a recombinant bacterium for an enzyme such as, say, human insulin?

First we would have to obtain the gene for human insulin, or a reasonable facsimile thereof; not a simple task. Essentially there are two ways to go about this: Burrow into the cell until we find the gene in question, or make our own.

Make our own? Is that possible?

Sure. The technology for making genes has been more or less available since Kornberg's work in the 1950s and 1960s. We'll look at some of the specific techniques available when we talk about gene machines later.

In order to make the gene, though, we have to have a thorough knowledge of its composition, and that may not be easy to come by. It would be easier, all around, to take an existing gene and go from there.

But even if we find the gene we want, it may not do us any good. Until now, we have talked about genes as though they are unbroken expanses of codon triplets which are precisely transcribed by mRNA and utilized in unaltered form for the constructions of proteins.

This is an accurate picture—of the genes in prokaryotic cells and in some of the simplest eukaryotic cells. But there are profound differences between the genes of single-celled organisms and the genes of men and beasts. For instance, while the prokaryotic genes are indeed unbroken expanses of codon triplets, the genes of higher organisms are filled with meaningless interruptions called *introns*, stretches of DNA that carry no genetic messages at all and which often take up considerably more space in the genes than do the coding sequences. Apparently (though the process is poorly understood), these noncoding sequences are transcribed literally by the mRNA but then are excised by some sort of editing enzyme before the mRNA can leave the nucleus.

The introns are a kind of garbage DNA, meaningless in terms of genetic information but significant for what they tell us about the evolution of the cell. There is considerable debate as to why the noncoding sequences exist at all. One theory holds that they are a kind of "selfish DNA," hitching a ride in the chromosome along with the more useful coding sequences, though they contribute nothing to the welfare of the organism.

But do the introns really need to contribute something to the welfare of the organism? The term "selfish DNA" is borrowed from evolution theory, specifically from a book by anthropologist Richard Dawkins entitled *The Selfish Gene,** which argues that DNA is its own excuse for being. Living organisms—including human beings —exist merely to enhance the gene's ability to survive. And if certain sequences of nucleotides find ways to survive that neither enhance nor detract from the survival of the rest of the organism, this is as valid an evolutionary gambit as any.

Be that as it may, noncoding sequences of DNA are so much garbage to the molecular biologist wishing to make a eukaryotic gene express itself in a prokaryotic organism. The bacterium into which the gene is spliced has no provisions for editing the introns out of the mRNA, so these noncoding sequences must be removed before the gene is spliced into the plasmid. A delicate bit of surgery indeed. How could this be done?

No one knows; the intron is a relatively new discovery and its ways still are mysterious. There is, however, an alternative approach to obtaining a copy of the gene that solves both the problems of the garbage DNA and of locating the gene.

The actual chromosomal gene is not required for gene splicing. All the potential gene splicer need do is find a cell that is actively producing the protein product of that gene and extract from that cell one of the mRNA copies of the gene being used in protein production; the mRNA copy will already have been edited by the cell's own enzymes. In the case of insulin, such copies are most likely to turn

*Dawkins, Richard. *The Selfish Gene*. New York: Oxford University Press, 1976.

up in the cells of the pancreas, where insulin normally is produced. Then a DNA copy of the mRNA copy can be produced in the laboratory through the use of the enzyme reverse transcriptase (which normally is used by RNA viruses that need to make DNA copies of their chromosomes to insert into bacterial cells).

Now at last we have a gene that can be spliced into a plasmid and inserted into a bacterium. But still this is not enough. Not only must the gene be inserted into the bacterium, but it also must be made to express itself while inside; otherwise all our work has gone for naught. The problem, once again, is the vast difference between prokaryotic and eukaryotic DNA.

Most one-celled organisms control the activity of their genes through the simple methods described in an earlier section—a sequence of nucleotides just above the gene serves as a mooring post for the enzyme that transcribes the gene into mRNA, provided that the mooring post isn't already occupied by a repressor protein. But the cells of higher animals utilize a different system of control, one that is not yet understood. There is some evidence that several sequences of nucleotides upstream from the gene serve as transcription initiators, but there also is reason to believe that the very shape of the chromosome—the way that the DNA is physically wrapped around the protein subunits that also form part of the chromosome—also is involved somehow in genetic control.

If the method of genetic control in the bacterium is so vastly different from that in the cell from which we took the gene we wish to splice into the bacterium—and if we understand so terribly little about these processes in the latter—how can we hope to manufacture human enzymes in bacteria?

One way might be to synthesize our own operator sequences and attach them to the gene that we splice into bacteria; this is the route that biologist Mark Ptashne of Harvard has been exploring. Ptashne has induced recombinantly spliced genes to express themselves in bacteria, using artificial control sequences. But the protein molecules produced by this process are immediately destroyed by the cell's own enzymes, and their amino acids are recycled for further protein production.

Apparently the cell does not recognize the recombinant protein as its own; it mistakes it for invading protein, perhaps of viral origin. One reason for this might be that natural proteins produced by the cell are given a special sugar coating—an escort of glucose molecules—which protects them from enzyme attack and identifies them as belonging to the cell. The recombinant molecules, lacking this identifying mark, are destroyed.

A better way to trigger expression of the gene is to take advantage of the cell's natural gene-control mechanisms. The recombinant gene can be attached to the plasmid in such a way that it is expressed along with one of the plasmid genes, and therefore its expression will be controlled by whatever controls the expression of the plasmid gene.

In one experiment this was accomplished by splicing the insulin gene immediately after the gene for beta galactosidase (which has a particularly well understood control mechanism). At the other end of the gene was placed a codon (either TGA or TAG) indicating where transcription was to cease.

Generally, several such recombinant plasmids are prepared at one time. They are then incubated together in a culture dish with *E. coli,* so that they can be "taken up" through the cell walls, becoming part of the bacteria. The bacteria then are encouraged to multiply. *E. coli* duplicates itself roughly once every twenty minutes; the recombinant plasmids replicate along with the bacteria, making copy after copy of the inserted insulin genes. This process is called *cloning* the genes.

Not every *E. coli* cell in the culture will contain a recombinant plasmid, of course; how can the ones that do be sorted out from the ones that don't?

You'll recall that in an earlier section we saw that bacterial plasmids often carry genes for antibiotic resistance. In preparing recombinant plasmids, biologists almost invariably use plasmids containing such genes; then, when the recombinant plasmid is introduced into a bacterium, that bacterium—and all of its descendants—are conferred with antibiotic resistance, while other bacteria in the culture are not. In order to see which bacteria received the plasmids, then, the cells are cultured in the presence of the proper antibiotic so only the recombinant bacteria survive. Thus the problem of sorting them from the nonrecombinant bacteria never presents itself.

The bacteria can be grown in any amounts necessary (though under earlier NIH regulations, such cultures were restricted to ten liters). In order to induce the recombinant gene to produce the protein for which it codes, the bacteria must be exposed to whatever molecule triggers the control mechanism of the plasmid gene to which the recombinant gene has been attached. In the case of the insulin gene appended to the beta galactosidase gene, the bacteria can be exposed to galactose sugar, which triggers the production of beta galactosidase. The beta galactosidase molecules produced by this process, however, have an artificial "tail" made out of insulin. The insulin molecules must then be clipped loose from the larger molecules.

How are they clipped? One way is to place a codon specifying the amino acid methionine at the point where the insulin gene is attached to its companion. The hybrid molecule thus produced then can be exposed to the chemical compound cyanogen bromide, which has the ability to clip protein molecules wherever methionine appears—and the insulin molecule will be cleaved from the beta galactosidase. The two molecules then can be sorted out by conventional procedures.

There are, of course, as many different methods of preparing recombinant proteins as there are scientists engaged in the research (multiplied, perhaps, by the number of different proteins being synthesized). The above description should give you some idea of the processes involved, though it is by no means an exhaustive catalog of the methods available.

The next several sections will look at where these techniques have brought us and at the industry that has grown up around recombinant DNA technology—an industry whose sole purpose is engineering the gene.

PART THREE

Selling Genes

THE ORGANISM INDUSTRY

How odd! An industry based on the reshaping of living organisms!

And yet the biotechnology industry is hardly new. It represents, instead, a reworking of several older industries—the pharmaceuticals industry, the chemicals industry, and perhaps the oldest industry of all: agriculture.

The new techniques of gene manipulation are just that: techniques. They can be used within a number of contexts, but mostly as means to a single end: the production of useful molecules. And the production of useful molecules by microorganisms is called fermentation.

Fermentation has been around a long time. Loosely put, it is a way of letting microorganisms do all the work of inducing bacteria and bacterial enzymes to turn out desired products. This sort of micromanufacturing has been available since long before recorded history and considerably before anyone knew that microorganisms were involved in the process at all. The ancient Babylonians and Sumerians used fermentation to produce alchoholic beverages, such as beer, more than 8,000 years ago. Needless to say, they were scarcely aware that they were drinking a molecular by-product of bacterial metabolic processes, or that they had millions of microscopic "slaves" producing their refreshments.

The microorganisms, in turn, had little reason to resent their enslavement; fermentation was something they did for their own sake. Many of the microorganisms involved in these processes were, and are, of a type known as anaerobic—organisms that do not (and cannot) use oxygen in their life processes.

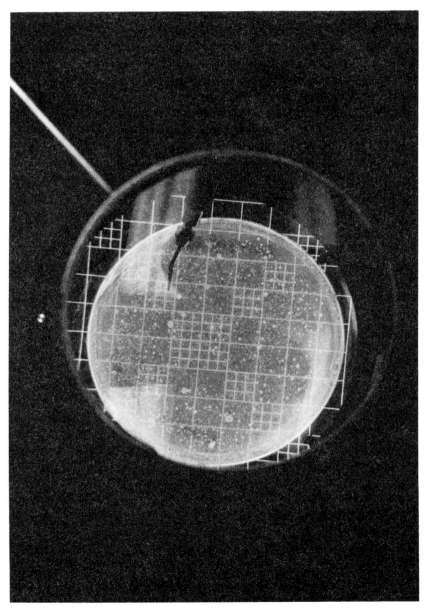

These small dots are recombinant colonies on a manual colony counter. These are the microscopic factories that are the basis of this new technology. *(courtesy Genex Corporation)*

The anaerobic organisms have a long and honorable history. Almost certainly the first living organisms were anaerobic, simply because there was no free oxygen on the primitive earth. (Considering the volatile nature of free oxygen and the fragility of the early organic molecules, this was probably for the best.) Though modern organisms derive energy from food through a process called respiration, in which the energy in the food is released by chemical reactions involving oxygen, anaerobic organisms lacked this option. Thus, for these organisms fermentation represented an alternative to respiration, a way of releasing energy without oxygen.

Fermentation, however, is highly inefficient compared to respiration, and organic materials processed by fermentation still contain a great deal of potential energy, much of which can be released by respiration; thus beers and wines, produced by fermentation, are rich with energy-carrying carbohydrates.

But we digress. The true cause of fermentation was pinpointed in 1857 by the great French chemist Louis Pasteur, who identified a number of living organisms involved in fermentation reactions. Before Pasteur's time, it had been thought that fermentation was a purely chemical process. (The very existence of microorganisms, in fact, had been unknown until about two centuries before Pasteur, when Dutch microscopist Anton van Leuwenhook turned his lens on a drop of water and discovered a world of microscopic creatures living within it.) Ironically, after Pasteur's time it was recognized that living organisms were not in fact required for fermentation; only their enzymes were necessary.

Enzymes are where you find them. Three thousand years ago, Homer made reference in the *Iliad* to a process by which the stomach of a lamb was immersed in goat's milk to make cheese. The process worked then and still works today, but the ancient Greeks were unaware that it was the enzymes contained in the stomach lining that produced the cheese.

Today we no longer need to throw the entire lamb's stomach into the milk to make cheese, nor do we necessarily need to use complete microorganisms to produce fermented beverages, though working with an entire microorganism does present certain advantages, especially if the chemical process consists of several steps. The proper microorganism can perform all or most of the steps itself, performing the task much more efficiently—and inexpensively—than any chemist. All we need supply is the proper conditions for the growth of the organisms and something for them to eat.

Gene splicing in most of its industrial applications is no more than a glorified fermentation process, a way of inducing useful molecules out of microorganisms, and so the technology that has been developed since ancient times to aid and abet in the fermentation process can be as easily used by gene splicers in the care and feeding of their recombinant organisms. What gene splicing offers that traditional fermentation processes did not is the ability to produce molecules for which no bacteria has ever had genes. In many cases, these molecules normally are found only in human cells.

The molecules produced by recombinant engineering are in turn useful in a number of other fermentation processes, those that use enzymes alone rather than complete microorganisms. Such processes generally involve placing the enzymes and a solution of the molecules to be acted on by those enzymes together in a single vat. After the desired chemical reactions take place, the enzymes are washed away in the purification process.

What a waste! One of the great values of enzymes is that they are not changed by the processes they help to promote. They are usable for an indefinite number of chemical reactions, yet in traditional fermentation reactions they are thrown away after a single batch of chemicals has been processed. For this reason we call the traditional method of fermentation the batch process. Because the enzymes for batch processing usually are hard to come by, and because they can be used only once, the products of fermentation usually have been rather expensive.

Now, however, modern technology offers us a better way of doing such things: continuous fermentation.

Continuous fermentation relies on the new technology of enzyme immobilization. Enzymes can be made to adhere to the surface of glass or plastic or ceramic beads, which in turn can be attached to solid support columns and placed inside fermentation vats. The action of the enzyme is unaffected by its immobilization—chemical processes, you'll recall, depend only on the shape of the enzyme, the way in which other molecules react to its surface; the enzyme itself does not become part of the reaction. A continous flow of raw materials can be forced through the vat to perform the desired reactions, but the enzymes need only be replaced every 750 hours or so.

The problem of enzyme cost is solved by recombinant DNA

technology. By engineering recombinant cells to produce the desired enzymes, an inexpensive and nearly limitless supply of enzymes can be made available for the fermentation process.

THE NEW PHARMACOLOGY

Though it was shown in 1897 that enzymes acting in a cell-free system (that is, outside of a living organism) could produce alcohol, the first industrial use of enzymatic reactions did not come until World War I, more than a decade later. Russian-born chemist Chaim Weizmann (later the first President of Israel) accidentally discovered a microorganism called *Clostridium acetobutylicum*, which produced several chemical compounds. Though no use for the organism was immediately seen (Weizmann had been looking for a way to produce synthetic rubber), it soon became apparent that the microorganism could be used to shore up Britain's dwindling wartime supply of acetone, a chemical used in the production of gunpowder. The production of acetone by bacteria became popular and remained the method of choice until the late 1940s, when it was superseded by cheaper processes involving petroleum, which in that bygone era was cheap and abundant.

The technological know-how gathered in the industrial fermentation of acetone proved its worth in the 1940s, when the production of antibiotics began in earnest and the pharmaceuticals industry adopted the methods of fermentation for its own.

The first antibiotic was discovered in 1928 by Alexander Fleming, who noticed that a certain variety of mold—*Penicillium notatum*—had the capability to kill certain bacteria he was studying in his laboratory. The substance that was secreted by the mold, dubbed penicillin, finally was isolated in the late 1930s by British scientists. It was first placed in large-scale production during World War II.

P. notatum was not the only microorganism to produce pharmacologically valuable molecules; within a few years, a full-scale antibiotics industry had come into existence based around a wide variety of microorganisms that had the convenient ability to produce molecules that could kill bacteria.

Just as the pharmaceuticals industry benefited greatly from the discovery of antibiotic-producing organisms, so it is the pharmaceuticals industry that stands to reap the most immediate benefits from recombinant-DNA technology.

In the pages that follow we will take a look at some of the pharmaceutical substances that the genetic engineers already have begun gearing up to produce.

Insulin: One of the smallest protein molecules produced by the human cell (fifty-one amino acids in two parallel strands), insulin also was the first such molecule to have its precise sequence of amino acids revealed—by English biochemist Frederick Sanger, in 1953. Because its sequence and form are so well understood, insulin lends itself readily to production through recombinant techniques.

Isulin is secreted by the beta cells, which are located in clumps of cells within the pancreas called the islets of Langerhans. Production of insulin is triggered by the presence of glucose—sugar molecules—in the body. (Apparently this is the result of the kind of negative feedback system studied by Jacob and Monod. The presence of glucose in the beta cells derepresses the insulin genes; the absence of glucose allows them to become repressed once again.)

The main purpose of insulin is to get the glucose out of the bloodstream and into the cells, where it can be broken down to release its energy and to power the activity of the cells; in this manner, insulin regulates the way in which the body metabolizes sugar.

Sometimes this system breaks down: The pancreas produces insufficient insulin and the glucose remains unmetabolized. This condition is called diabetes. Left untreated, it can be fatal.

Fortunately, diabetes is easy to detect (by the presence of excess sugar in the blood), and, while there is no actual cure, there is a simple treatment: periodic injections of insulin. But, while this restores to the body a balanced sugar metabolism, it leaves the diabetic open to certain complications and side effects, such as progressive tissue damage and blindness.

This is due, in part, to the use of nonhuman insulin in injections for diabetes. The insulin used for treating diabetics is generally taken from either pigs or cows; human insulin simply isn't available in the necessary quantities. Pig insulin and cow insulin bear a fortuitous resemblance to the human kind, but they are not identical; in each case, there is a single amino acid difference. For most purposes this difference is meaningless; pig and cow insulin regulate human glucose metabolism quite effectively. But they can't deal with the side effects, and some few diabetics simply are allergic to the nonhuman insulin.

With recombinant-DNA techniques, insulin has been produced that is identical to the human kind in every essential way; and, because it does not have to be filtered out of the bodies of slaughtered animals, recombinant insulin should be less costly to produce than other kinds.

In almost every way, human insulin is an ideal target for recombinant production and probably will be one of the first such products available commercially.

Interferon: If there is a glamor chemical in the gene-splicing revolution, surely it is interferon. Touted as the all-purpose medicine, effective against everything from viruses to cancer, interferon until now has been available in no more than minute quantities, making these claims somewhat hard to prove (or to disprove). With recombinant technology, large quantities of interferon soon should be available and significant experimental data easier to come by.

Discovered in 1957 by Alick Isaacs and Jean Lindemann of Mill Hill Hospital in London, interferon is produced naturally, but grudgingly, by the cells of the human body. It is an unusually large protein molecule and it comes in three known varieties: leukocyte, or alpha, interferon; fibroblast, or beta, interferon; and immune, or T interferon.

Interferon is part of the body's immune system; specifically, it is one of the body's weapons against viruses. When a cell is invaded by viral DNA, the interferon gene is activated, and interferon molecules are produced. The interferon molecule makes no attempt to fight the invading viruses; instead, it flees the cell altogether and enters the intercellular fluid, eventually attaching itself to a receptor molecule on the surface of a nearby cell.

From this vantage point, the interferon molecule somehow stimulates the new cell to produce antiviral enzymes; in effect, it warns the cell that a virus attack is imminent and encourages it to mount a counterattack. When the viruses arrive, the cell attacks the viral DNA with restriction enzymes that scissor it into fragments and with enzymes that block its ability to use the cell's reproductive machinery to create new viruses.

The very presence of interferon molecules in the body signals to special cells called *"natural killer" (NK) cells* that their services are needed. The NK cells speed to the site of the invasion and attack the viruses, having been alerted to their presence by the interferon.

Interferon, then, is an essential part of the body's antiviral defense. It follows that interferon may have certain value in the treatment of viruses, and experiments in the 1960s showed this to be so: Interferon was in fact a powerful antiviral medicine.

But this information was useless, for the simple reason that interferon was almost impossible to produce. In order to isolate a mere 100 milligrams of the substance, the molecules had to be filtered out of 65,000 pints of human blood. The resulting interferon preparation was still far from pure—less, in fact, than 1 percent actual interferon—and exorbitantly expensive. Because of this expense, interferon research languished until the early 1970s.

What spurred new interest in the substance was the felicitous discovery that it had anticancer properties. Subjects with certain forms of cancer showed, in some experiments, a strong improvement when given interferon in regular doses. Further, interferon had certain natural advantages as a cancer drug: It could be used concurrently with other, more traditional methods of chemotherapy and, unlike many other forms of cancer treatment, it did not leave the patient vulnerable to viral infection and therefore to serious complications. (Many, perhaps most, cancer patients die not of the disease itself but of complications brought on by their weakened conditions.) In fact, the interferon actually increases the patient's ability to deal with infections.

Alas, in other experiments the results were more ambiguous; there was just not enough interferon available to mount a proper battery of experiments. And what there was of it was so diluted as to be nearly worthless.

Now, however, we can isolate the genes for various types of interferon and place them in recombinant organisms. This offers the possibility of a nearly endless supply of the substance for experimental purposes; whatever uses it may have in the war against cancer now can be fully exploited. Furthermore, when interferon is in adequate supply, other avenues will open to exploration. It will be possible to examine its effectiveness as an antiviral drug. Oddly, it also may be used as a method for suppressing the immune system rather than stimulating it. When interferon is administered very early during the response to a viral attack, it has the peculiar effect of reducing the immune response. Only when it is introduced later does it stimulate it. Thus, for example, interferon might be used to prevent rejection of transplanted organs.

There is some concern, however, that extended interferon therapy might prove too much of a good thing. The body produces interferon only intermittently, and in small amounts. It is flushed quickly from the system; the day after it is produced, it is gone. The body may have its reasons for this interferon stinginess: Prolonged exposure to the substance may be hazardous to one's health. There may be side effects of which we are not yet aware. Only time and further research will tell.

Human growth hormone: A moderate-sized protein— roughly 195 amino acids in length—growth hormone is secreted by the pituitary gland. As its name implies, it controls the process by which the human body grows, particularly in the years leading up to adolescence.

When the child's production of growth hormone malfunctions, the result is a condition known as pituitary dwarfism. Apparently this condition results not from an absence of the hormone but from a defect in the gene that produces it, so that the molecule itself is malformed and therefore useless. Victims of this condition retain childish facial features and doll-like (though normally proportioned) bodies into adulthood. But unlike other forms of dwarfism, the condition responds to injections of growth hormone, and a child treated early enough in life can grow to normal proportions.

As with interferon, however, human growth hormone is expensive and difficult to come by (though it is generally offered free to victims); it can be obtained only from the pituitary glands of corpses, and fifty such glands are needed to treat one child for a year. So that there will be enough for all sufferers, an individual child can be treated only until the low range of normal height—about five feet— is reached.

Also like interferon, growth hormone is species-specific, so that animal growth hormones cannot be substituted.

Recombinant production of growth hormone will not only allow further treatment for each child, but it also would make possible research into other anticipated benefits of the hormone—in healing burn wounds and ulcers, for instance.

Synthetic vaccines: When a foreign particle—a virus, for instance—enters the body, it is swiftly targeted for destruction by the body's immune system. Molecular structures called antibodies, tailored to recognize specific invading particles, will attack these particles and disable them. The particles that the antibodies are tailored to recognize are called antigens. For each different antigen, the body forms a unique antibody.

When the body is first exposed to a particular antigen—a specific strain of virus, say—the production of antibodies is sluggish; the specific antibody for that particular antigen must be designed before it can be utilized in the attack against the invading particle. Upon repeated exposure to the antigen, however, the body "remembers" the recipe for the needed antibody and attacks the antigen before any ill effects can ensue. This explains why we develop immunity to most diseases after a single exposure; the second time, our bodies have but to extract the proper antibody from the repertoire built up during previous contacts.

As ingenious as the antibody-antigen response may be, it still leaves the body open to debilitating (and perhaps fatal) attacks by disease-causing organisms against which no antibody has yet been formed; the time lag between the onset of the disease and the counterattack by the immune system is a dangerous period, and anything that shortens it is desirable.

The process of vaccination is designed to close this window of vulnerability by stimulating the body to create antibodies against diseases it has not yet been exposed to.

How is this done? Antibodies recognize antigens by the shapes of certain molecules on their surfaces; if these molecules are present, the antibody attacks, whether the antigen represents any clear danger to the organism or not. Even if the antigen is dead, the shape of its surface is sufficient to stimulate the immune system. Hence, the body will as readily produce antibodies to a dead or weakened virus as to a live one. All that matters is the shape of the viral coat.

In the past, the chief method of producing a vaccine has been either to kill the virus in question (by exposing it to high temperatures, for instance) or to breed a weakened version of it (by culturing the virus in the bodies of certain animals or in the yolks of eggs). When injected into the human body, the dead or weakened viruses stimulate the production of antibodies without causing the disease, preparing the body in advance for an attack by an unweakened strain.

Vaccination has its drawbacks, however. Working with an actual virus is dangerous, and even a weakened virus should be injected into a patient with great care. On occasion, a vaccine still may turn out to be virulent and actually can cause the disease it is intended to prevent.

One promising alternative to the traditional vaccine process is the synthetic vaccine; because it does not involve the use of actual viruses, there is no risk of accidental infection. Because antibodies recognize antigens by small particles on their surfaces, it is possible, through gene synthesis and recombinant-DNA technology, to design a bacterium that will produce protein molecules identical in size and form to these particles. These "artificial antigens" then can be injected into the body to stimulate the production of those antibodies. The process is the same as in normal vaccination but with none of the potential side effects.

There is a problem with this method, but it is a problem shared by killed-virus vaccines: The amount of antigen injected into the body is not great enough to produce large amounts of antibodies. When a living virus (weakened or virulent) enters the body, it is able to reproduce continually, supplying the immune system with a continuous supply of new antigens on which to model its antibodies. A greater resistance is developed; hence live-virus vaccines are more effective, albeit also more dangerous, than killed-virus vaccines.

It may be possible, however, to inject not synthetic antigens into a patient but the actual microorganism that produces the synthetic antigens, which then can continue to multiply and produce the antigens within the body itself, continually stimulating the immune system to produce antibodies. Whether this would be a safe and effective form of vaccine, though, remains to be seen.

Monoclonal antibodies: Though not produced by recombinant-DNA techniques, monoclonal antibodies represent one of the most exciting and potentially useful products of modern genetic technology.

There are a number of medical uses for antibodies: as supplements to one's own natural antibodies, for instance, or for detecting the presence of antigens in the body (which is an effective way of diagnosing a disease). In order to obtain these antibodies, they must be purified out of human blood at great expense and effort, but antibodies obtained by such methods rarely are very pure. Nonetheless, they are extremely expensive.

Antibodies are produced by white blood cells: Each cell (and its descendants) produces a particular kind of antibody, and only that antibody. If somehow we could culture the white blood cells that produce antibodies—if we could grow large quantities of them in fermentation vats, as we grow certain strains of bacteria—we would have an unlimited supply of pure enzymes; if we started with but a single white blood cell, all of the cells in the culture (each a direct descendant of the original) would produce the antibody that the original cell had been programmed to produce, and no other. Purification would be unnecessary.

But this cannot be done. Human body cells cannot be cultured as bacterial cells are, because each cell can, on the average, duplicate itself only fifty times. (This is called Hayflick's limit, and it is true of human body cells in general.) A human cell can produce only so many offspring; then the entire cell line dies. For this reason, human body cells cannot be cultured.

There is an exception to this rule, however. There is one type of human body cell that apparently can undergo an infinite number of divisions, in defiance of Hayflick's limit: cancer cells. This, in fact, is why cancer cells are undesirable—they don't know when to stop. A cancerous tissue will grow until it destroys the organ of which it is part, and it can send out tiny seeds (called metastases) through the rest of the body, where they can act in the same way.

What we need is a cell that can produce antibodies like a blood cell and yet can multiply indefinitely like a cancer cell. Such hybrid cells could produce pure antibodies in industrial quantities, for a reasonable cost. Perhaps we could design such a cell using recombinant-DNA techniques—building hybrid cells is what gene splicing is all about, after all—but our knowledge still is too limited. We have no idea as yet what combination of genes sends a cell into cancerous reproduction, nor how to inject the genes for producing antibodies into a cell already turned malignant.

There is another route that we can take, however: cell fusion. In this process, two cells can be brought together in a laboratory dish and treated with chemicals that partially dissolve the cell membranes. The half-naked cells then can meld together, their cell walls oozing into a single wall; the two cells become one. Because the resulting hybrid has a double complement of genes, the cell is forced to slough off a portion of them.

Cell fusion is a kind of biological crap game—the biologist has no control over what genes are retained by the hybrid cell and what genes are discarded. He can only toss the dice and hope for the best.

In 1975, when Georges Kohler and Cesar Milstein of Cambridge University fused blood cells from mice with cancerous mouse cells, they rolled a lucky seven. The resulting hybridoma—the term used for a cell combining the ancestries of a normal cell and a cancerous cell—retained both the ability to produce antibodies and the ability to multiply indefinitely.

Thus these hybrid mouse cells are capable of producing a nearly unlimited supply of pure antibodies. Because each hybridoma culture produces a single line of identical antibodies, their product has been termed monoclonal antibodies. What are some of the uses to which we can put these monoclonals?

They are powerful diagnostic tools. Because antibodies are designed to recognize and attach themselves to specific antigens, they tend to clump together in that part of the body where the antigens are present. Injected into the body of a cancer patient, monoclonals tailored for the detection of tumors will zero in on the cancer like magic bullets. If the monoclonals are synthesized from radioactive molecules, their progress through the body can be traced with special instruments and a map of the cancer's progress devised. Other diseases could be similarly mapped, using monoclonals specific for those diseases.

Antibodies could even carry medicine into the body, delivering it directly to the site of the infection. One of the drawbacks of cancer chemotherapy is that the chemicals which destroy the tumors often destroy healthy tissue as well; with monoclonals carrying the chemicals straight to the tumor, surrounding tissues would be unaffected.

Monoclonals also could be used for tissue typing. Antibodies, as noted in an earlier chapter, are able to recognize the specific patterns that identify a certain tissue as part of the "self"—that is, that the tissue belongs in the body and is not foreign matter. This identification is facilitated by the existence of special recognition sites on the surface of the body's cells, tiny structures called HLA antigens. On every cell of every tissue in every human body, there are four of these HLA antigens, and they are identical on every cell in the body. Yet they differ from body to body, like fingerprints. Thus each cell is identified as to what body it belongs in. The cells in one body would not be acceptable to antibodies designed to recognize cells from another person's body; the HLA antigens would be different.

Each of the four antigens sits in its individual position on the surface of the cell; for each of these positions, there are only a certain number of possible antigen structures, and these structures can be identified by number. The specific pattern of HLA antigens on an individual's cells therefore can be given an identifying sequence of numbers, depending on which specific structures are present, just as an individual's blood can be given an identifying letter as to its type, and just as an individual's fingerprints can be reduced to a written description of their whorls and lines.

There are more than a million possible combinations of HLA antigens—a large number but not an infinite one. This becomes significant when, for some reason, tissues from one body must be introduced into another: during an organ transplant, for instance. The immune system rejects tissues that don't have the proper HLA signature. For this reason, transplanted organs often are taken from a close relative of the patient, because with genetic closeness there is an increased chance that the HLA signatures on the tissue will be similar or identical. But the finite number of possible antigen combinations means that probably there will be other individuals with identical or similar HLA signatures on their tissues as well; if a relative is not available, perhaps a stranger can be found with a compatible tissue type.

But how to find these people? Monoclonals could be used to identify specific HLA structures, a process that, until now, could be done only with great difficulty. With monoclonals, tissue types can be diagnosed as readily as blood types. Conceivably, banks of information could be kept on individual HLA numbers, and when an organ is needed for transplant these banks could be consulted. Or, if an organ should become available for transplantation, it could be typed and routed wherever it was needed.

Monoclonals are useful molecules. They also could be used for pregnancy tests and drug detection—the list is nearly endless.

What Kohler and Milstein produced in their 1975 experiment was a fusion of mouse cells; obviously, human antibodies would be more valuable for many of the applications suggested above. Human hybridomas have proven harder to come by, but that breakthrough already has been achieved by several different laboratories, and before long, human antibodies should be available in quantity.

Neuropeptides: The activity of the brain, like that of any other organ in the body, is controlled by hormones. Called neuropeptides, these hormones have a profound influence over our thoughts and emotions; they affect the way we feel and perceive;

they are responsible for hunger, desire, tranquillity, and anxiety; and they can be synthesized using recombinant-DNA techniques. One minor brain hormone, somatostatin, already has been cloned in bacteria. Though its role in the brain is poorly understood, somatostatin may be in charge of regulating other hormones throughout the body.

Little is known about how neuropeptides in general perform their functions, or what those functions are. We do know that there are neuropeptides—beta endorphin, for instance, and the enkephalins—which are similar in structure to morphine and have much the same effect on the brain. In fact, it probably is morphine's resemblance to these natural brain opiates that allows it to function as it does. If we can isolate these natural chemicals they may provide us with a safe, nonaddictive substitute for a number of dangerous drugs.

One recently discovered neuropeptide, dynorphin, shows promise as an extremely effective pain-killer. Another, lutenizing releasing hormone (LRH), may turn out to be a genuine aphrodisiac.

By controlling these neuropeptides, it may be possible to control a person's sleeping and eating habits, their ability to learn and recall and feel pain, even their sex drive. The neuropeptides may prove to be a treasure trove of psychopharmacological riches, if only we can produce them in sufficient amounts through gene splicing to allow research in these chemicals to proceed apace.

This, then, is the new pharmacology, a medicine chest full of the body's own chemicals, manufactured by tiny cellular factories that can be brewed by the millions in huge vats, where a fermentation technology older than written history meets a biochemical technology still in its rambunctious infancy. But the possibilities inherent in the new genetics extend far beyond the pharmaceuticals industry.

CELLULAR FACTORIES

The chemicals industry, which synthesized certain of its products biologically long before the first bacterially produced drugs were marketed, stands to gain considerably from this technology, especially in light of the rising cost of petroleum.

In a way, this is ironic. The chemicals industry discovered the value of bacteria when Weizmann began producing acetone from microorganisms during World War I, but by the late 1940s it had proven cheaper to derive chemicals from the breakdown of petroleum into its component organic molecules—to produce chemicals through strictly chemical processes.

The reason was largely economic. Petroleum has traditionally been cheap and widely available. Bacteria, on the other hand, must be fed expensive raw materials, such as molasses and starch. Making chemicals from petroleum was simply cheaper.

No longer. With the oil shortages of the 1970s, the price of oil has risen sufficiently that competing technologies have become attractive. Furthermore, chemical synthesis produces a wide variety of waste products, which have to be disposed of—and waste disposal costs money. Biological processes, on the other hand, utilize most of the by-products of the synthesis and therefore produce little waste. And because there are few waste products, biosynthesis creates little or no pollution, an important consideration in an age of expensive pollution controls (though, ironically, the laws that encourage such pollution controls are being eased even as biotechnology offers a relatively inexpensive way of complying with them).

Another advantage of biosynthesis is that an operation that might take many steps with conventional chemical production methods often can be performed entirely within a single microorganism; the bacterium, in essence, does all the work. And because the synthesis is performed with enzymes, it takes place at moderate temperatures and pressures, unlike many chemical processes.

Obviously, recombinant DNA technology is not required for many forms of biosynthesis—Weizmann's acetone-producing bacterium, for instance, was a naturally occurring organism. In the past, when biologists required an organism that did not occur conveniently in nature, they would breed bacteria in quest of desirable mutations which, when they appeared, were encouraged to multiply.

With gene splicing, it will be possible to produce through biosynthesis any organic molecule that exists in nature—providing we can locate and analyze either the molecule or the gene that codes for that molecule. We may even be able to produce a few molecules that do not exist in nature. Our only limitation is knowledge—what we understand about the nature of molecules and the nature of genes. Unfortunately, that limitation is, at present, a fairly severe one: For all that we know, our ignorance still is considerably vaster. But the gap is closing rapidly. Our knowledge of genes and molecules is expanding at an almost frightening rate.

One area in which biosynthesis eventually may prove essential is in the production of fuels. All fossil fuels, such as oil and coal, are biologically derived. Plants and other organisms, which gain their energy directly from the sun through a process called photosynthesis, are imprisoned beneath the earth's crust for millions of years, where they are reduced to a highly concentrated form. We can use these ancient organic molecules to power our automobiles and factories, but the supply, while large, ultimately is limited.

We can, however, produce alcohol through biosynthesis and use that alcohol to extend our vanishing fuel supplies in the form of gasohol, a combination of gasoline and alcohol. The problem is that, even in an age of rising energy costs, alcohol still costs more than petroleum, though this is not likely to remain true for long. Another drawback is that, using current distilling processes, it takes more energy to create alcohol in a usable form than the amount of energy the alcohol will produce when used as a fuel.

Gene splicing, however, gives us the ability to create custom-made microorganisms which can produce higher yields of alcohol. Because the raw materials for this sort of biological conversion often are foodstuffs such as corn, there may be some conflict in diverting these materials from other uses; but the portion of the corn that is extracted and used to create alcohol is relatively small, and what remains of the corn used for this purpose could be recycled for animal feed, thus allowing us to produce large amounts of alcohol while only slightly reducing the amount of foodstuffs available—altogether a more efficient use of the materials involved.

Microorganisms also may prove useful in mining fuels. Bacteria exist that produce substances that can be sprayed into supposedly dry oil wells to extract oil that would otherwise be uneconomical to remove. Other microorganisms are under development that would draw oil out of shale and tar sands, thus opening up huge reserves of fossil fuels not currently available at reasonable cost.

Similarly, bacteria could be designed to enter mines and leach pure metals from metal ore, so that the pure metal can be flushed out of the mines with water.

There are any number of organisms that break down pollutants into their harmless components—bacteria, for instance, that can

How to get what you want out of DNA.   Once you've decided what you want to produce, you obtain the desired gene
from (1) a gene machine or (2) a biological source such as
tissue. Now you have a gene (3) containing the genetic information to "code" a cell or organism to produce a substance
like human interferon or human insulin. (4) Control signals that
will attach the gene to its carrier are added on. While you've
been isolating or assembling the gene, molecules called
plasmids (6) have been pulled out of a microorganism such as
E. coli. The plasmids are cut open (7); then the gene and the
plasmid are spliced together (5). This produces a recombinant
DNA molecule (8). This molecule is then introduced into a host
cell (9).

Once a plasmid is inserted into a host cell, it will be copied
and recopied many times (10); as it is copied, the cell will act
on the instructions that the gene contains—it will produce the
insulin or interferon or whatever—in a process called expression (11). Even as it does this, the cell will divide and produce
offspring, which will produce offspring, which will continue this
process in turn, because they will all contain the new gene
(12).

Growing or *fermenting* a lot of these genetically engineered
microorganisms is first done in shaker flasks (13) in an effort to
figure out the best way to keep these little guys happy. Then,
the process is scaled up (14) until (15) you're producing
enough stuff to sell. At this point, you (16 and 17) extract and
purify the substance you started after in the first place, pack-
age it, and try to find a market. If you're making an industrial
chemical like alcohol, all you have to do is sell it (18); however,
if you're making a drug or other health-care product, there are
additional steps to be taken.

First, you test your product on animals to show how well it
works and how safe it is (19). Next, you present it to the Food
and Drug Administration; if they think it's okay they'll let you try
it on people. Once you're satisfied that the product works
on people and that it *is* safe, you file a new drug application
(NDA) with the FDA. When they approve that, you can finally
market your product in the United States.

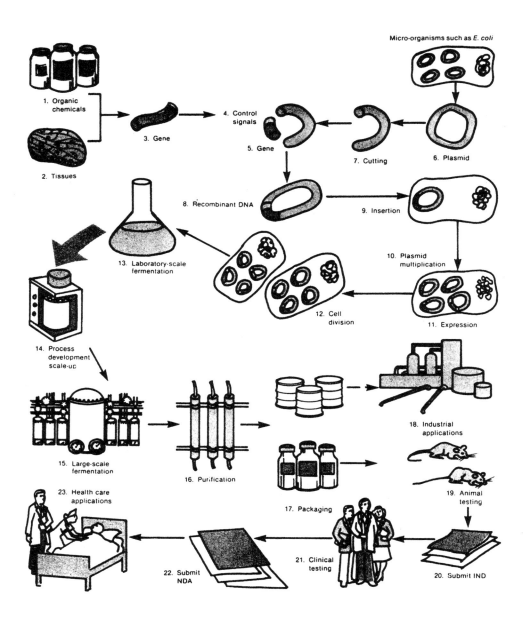

Micro-organisms such as *E. coli*

1. Organic chemicals

2. Tissues

3. Gene

4. Control signals

5. Gene

6. Plasmid

7. Cutting

8. Recombinant DNA

9. Insertion

10. Plasmid multiplication

11. Expression

12. Cell division

13. Laboratory-scale fermentation

14. Process development scale-up

15. Large-scale fermentation

16. Purification

17. Packaging

18. Industrial applications

19. Animal testing

20. Submit IND

21. Clinical testing

22. Submit NDA

23. Health care applications

degrade the oil in oil slicks. Bacteria also have been engineered to eat insecticide residues and other toxins, thus cleaning the environment.

The food-processing industry also should profit from the new genetic technology. Organisms can be designed that turn inedible substances such as cellulose into nutritious protein that can be used as animal feed or even as food for human beings. Useful nutrients can be derived from this so-called single-celled protein (SCP) either by removing the amino acids from the organism and using it as a food additive or by processing the organisms themselves to be used as food. Though the idea of eating bacteria for breakfast may sound less than appealing, clever food processing can make SCP taste just like meat (or a reasonable approximation thereof). In times of food shortage, and in parts of the world given to perennial famine, SCP may prove a dietary godsend.

GREEN GENES

In the long run, the field most profoundly affected by the new genetics technology probably will be agriculture—a field that already deals in the very stuff of life. Scientists and farmers have worked together for years—improving fertilizers, improving growing conditions, improving methods of harvesting—but it is to the science of genetics that agriculture always has been most closely wedded.

The farmer who carefully selects the seeds of those plants that best meet his specific needs and favors them in the next year's planting is performing a rudimentary sort of genetic experiment; in effect, he is taking evolution in his own hands and redesigning the natural genetic tendencies of those plants to his own ends. If a certain variety of a certain plant produces a greater marketable yield than other varieties of that plant, then the farmer will favor that variety; in time, it will come to outnumber its less prolific siblings, simply because it will be planted more often. Similarly, if a particular breed of cow gives more milk, or produces more beef, than another breed, the offspring of that breed will be favored above others. This sort of forced evolution generally is referred to as artificial selection, as opposed to the more common evolutionary process of natural selection. The difference is that in the latter it is the blind forces of nature doing the selecting; in artificial selection, it is the farmer.

Artificial selection, though faster than natural selection, is nonetheless a slow process. It can produce profound changes in living organisms, but only over many generations. The process can be accelerated, in some instances, by exposing the seeds of plants, say, to radiation, encouraging the genes to mutate so that new and desirable characteristics can appear and then be selected for, but even this scattershot process is restrained by chance and the limitations of nature. It is unlikely to produce superplants; only better ones. And, of course, we hardly could use this method on cows.

Genetic engineering offers a better alternative. If it were possible to reach into the sex cells of a plant or even an animal and rearrange the genes by whim, then an almost limitless number of genetic alternatives could be explored. Crop yields might be doubled, tripled, even quadrupled. Plants could be developed that drew essential nutrients from the air itself. Perhaps we even could develop cows that give chocolate milk.

But is this within the reach of current technology? Rearranging the genes of a plant is a far cry from inserting recombinant plasmids into *E. coli.* Concocting supercows is a quantum leap beyond brewing bacteria in a fermentation vat.

And, in fact, the cow that makes chocolate milk is only a dream at this point (though we will explore the possibilities of mammalian genetic engineering in Part IV), but soon we may have the ability to tamper with the genes of plants much as we now tamper with those of *E. coli.*

How? Well, there are several possible ways in which we can dabble in plant genetics. We might, for instance, use microinjection (whereby molecules can be inserted forcibly and directly into living cells) to install the desired genes in the plant. But experiments in this sort of genetic surgery have yielded only poor results; we must look elsewhere for the key that will admit us to the plant gene.

That key, most likely, will be a microorganism. Just as viruses prey on animal and bacterial cells by injecting their DNA into unsuspecting hosts, so there are organisms that are capable of inserting their genes into the cells of plants. If the gene we wanted to insert into a plant could first be inserted into one of these organisms, perhaps it too could be insinuated into the plant genome, riding in with the invading DNA just as recombinant genes ride into bacteria on plasmids.

There are, for instance, the *viroids*, tiny loops of naked DNA twenty times smaller than the smallest virus, so small that a viroid next to a plasmid would look, in the words of one researcher, "like a Cheerio next to a hula hoop." Though the viroids possess only 300 nucleotides and no identifiable protein genes, somehow they are capable of inserting their genes into plants as large as trees and infecting them with destructive diseases, perhaps by somehow imitating and confusing the control signals that regulate the plant's own genes.

Could genes be spliced into viroids and inserted thereby into plants? Perhaps; but at this point we understand too little about what viroids are or how they work to have much hope of using them as vectors for genetic engineering.

A better possibility is the cauliflower mosaic virus. As its name implies, CMV is a virus that infects cauliflowers and related plants; researchers have already managed to splice recombinant genes into the CMV genome and insinuate those genes into plants, but none have yet been coaxed into expressing themselves. CMV, however, may prove very useful in the future.

The best bet for plant genetic engineering right now is a bacterium called *Agrobacterium tumefaciens*—Agrobacter, for short. Altogether an amazing bacterium, Agrobacter is the only organism we are currently aware of that is designed for the pupose of injecting prokaryotic genes into eukaryotic cells—exactly what the gene splicers ordered. Contained within Agrobacter is a tiny loop of DNA called the Ti plasmid, made up of some 200 genes. The bacterium actually has the capability of attaching itself to a plant cell and injecting its plasmid into the nucleus, where a portion of the plasmid ring twelve genes long, called tDNA (for transfer DNA), jumps out of the plasmid and incorporates itself into the plant cell chromosome.

The tDNA genes induce the plant to produce certain nutrients required by Agrobacter; they also create a tumorous swelling called a crown gall on the side of the plant.

We know that it is possible to slip recombinant genes into plants via Agrobacter simply because it has been done. Researchers have spliced several protein genes into plants by this route and they seem to be able to express themselves. (In one experiment, genes from a bean plant were inserted into a sunflower. The resulting hybrid was nicknamed a "sunbean.")

One drawback of gene transfer by Agrobacter is that the bacterium is fairly choosy about what plants it will infect; the cereal crops, which are considered valuable targets for genetic engineering, are immune to its intrusions. Furthermore, no one is sure that genes injected via Agrobacter can be made to express themselves in future generations of the treated plant, which is an important consideration. A recombinant gene that cannot be carried from generation to generation is of only limited use.

Given, however, that we have the ability to perform genetic engineering on plants, what sort of traits would we want to engineer?

We might design a plant that uses photosynthesis more efficiently, drawing more energy from a given amount of sunlight. Such plants would produce a greater yield for less effort and therefore would be intrinsically valuable.

We might want to create a plant that would be tolerant of salt; such plants could live in desert regions where the soil is heavily saline, where other plants would quickly die. We know that genes for this trait exist, because there are plants that express it, but it might prove valuable to spread this trait around.

At the moment, the pet project of most gene splicers in agriculture is to develop either a plant or a plant-bacteria combination that is capable of deriving its own nitrogen from the air. To understand why that would be important, let's take a look at the process called *nitrogen fixation.*

The element nitrogen is an essential part of all protein molecules and therefore an essential part of all living creatures. It is also the most abundant element in the earth's atmosphere: more than 70 percent of the air we breathe is nitrogen.

Yet, unlike many elements, nitrogen atoms are slow to react with other atoms and therefore are extremely reluctant to become parts of molecules; under normal conditions, they simply won't bond. How, then, does nitrogen become incorporated into protein?

Because molecular nitrogen is referred to as "fixed" nitrogen, the process by which it is made part of a molecule is called nitrogen fixation; this process is mediated by the enzyme nitrogenase. Although nitrogenase can be found in bacteria, it almost never occurs in higher organisms. Human beings and other animals don't need it, of course; they get all the protein molecules they need from plants and other animals. But where do plants get the nitrogen for their own protein?

They get it mostly from bacteria, either directly or indirectly. Legumes—a type of plant which includes peas, alfalfa, and clover—have developed through millions of years of evolution a symbiotic relationship with a bacterium called rhizobium, which possesses the nitrogen-fixing gene. The bacteria of rhizobium live in a tiny bump, called a nodule, on the surface of the plant and receive nourishment from their symbiotic partner; the plants, in turn, receive fixed nitrogen from the bacteria, which they can then easily incorporate into protein molecules.

The relationship is a convenient one, and not just for the legumes. When the legume dies it leaves a residue of nitrogen in the soil, which then nourishes other plants that grow in the same place at a later time; this is why farmers often plant legumes in a given field during alternate seasons—to refresh the nitrogen in the soil.

Alas, this nitrogen often is leached from the soil by rain and other natural processes, so farmers are forced to use nitrogen fertilizer to supplement the fixed nitrogen supply of their nonleguminous plants. Nitrogen fertilizer is expensive—farmers use over one billion dollars' worth a year—and making this fertilizer consumes a great deal of energy. But without it, modern intensive farming would not be possible.

Yet with genetic engineering it is conceivable that the gene for nitrogenase could be spliced into nonleguminous plants so that they could derive their fixed nitrogen directly from the air.

This would be quite a boon, because it would eliminate the need for expensive nitrogen fertilizers, but whether it is feasible is anybody's guess. Genetic engineers have successfully designed plasmids containing all of the genes necessary for nitrogen fixation, but they have not yet induced them to express themselves in a plant. One problem with this system is that nitrogenase is extremely sensitive to oxygen; it cannot function in its presence. In the legume-bacteria relationship, complicated arrangements have been made to protect the nitrogen-fixing enzyme from exposure to oxygen. Can this be duplicated in genetically engineered, nitrogen-fixing plants?

Maybe; then again, maybe not. An even more daunting problem is that the nitrogen-fixation process requires large amounts of energy, which will have to be supplied by photosynthesis. Because each plant is equipped only to supply the amount of energy needed for its normal metabolic processes, nitrogen fixation will have to be performed at the expense of other processes, such as growth.

Nitrogen-fixing plants might give low yields, which would cancel out the value of their unusual gene.

A more valuable approach might be to design nitrogen-fixing bacteria that can live in symbiosis with nonleguminous plants just as rhizobium lives with legumes. But this might involve the genetic engineering of the plants as well as the bacteria. It takes two to make a symbiosis; the plant will have to learn to play its own part in the relationship.

Whatever the outcome of this particular experiment, the future looks good for gene splicing in agriculture, though most of the major victories still may lie some distance in the future. As populations expand and available agricultural land shrinks, world food crises may lie around the corner. Genetic engineering may turn out to be a necessary step in the future of agriculture, and it could be arriving just in time.

THE BUSINESS OF BIOENGINEERING

Recombinant DNA means many things to many people. To the scientists who devised the techniques it is a means toward an end: understanding the complex workings of the genetic mechanism. To the entrepreneurs who intend to manufacture the biological products that recombinant DNA makes feasible (many of whom, ironically, are the same scientists who devised the techniques), it represents something far more tangible: money.

Bioengineering is big business—at least, it is on the verge of becoming big. Hundreds of small companies sprang up during the 1970s to take advantage of its potential, some of them vanishing as suddenly as they had arrived. Even the large, established chemical and biological firms, founded on earlier methods of producing these products, deigned at least to notice this upstart technology in their midst, though most remained cautiously aloof from the action, allowing the young companies to fight out the technical and commercial details among themselves, perhaps to swoop down like hawks at a later date when things have sorted themselves out.

The idea of an industry that has the power to alter life is a fascinating one, and it raises important questions. Does any corporation have the right to deal in living organisms? If an industrial scientist invents a bacterium that produces a marketable product, can that scientist and the corporation that he works for claim exclusive jurisdiction over that organism and its valuable functions?

Can living things be patented?

Whether or not corporations have the moral right to do such things is arguable. As our knowledge of organic processes has grown, the distinction between life and nonlife has grown ever vaguer; if we view a living cell as a kind of molecular machine, a device put together by nature (or ingenious biologists) to draw energy from the environment and reproduce its kind, then surely anyone who builds an improved device of this type should have proprietary rights in it. But if we view life as something transcending the simply mechanical, something separate and apart from the inorganic, then the question becomes more complex. As Dr. Harold J. Morowitz, author of *The Wine of Life,* * puts it, the whole business of corporate gene splicing and patenting living organisms is nothing more than "reducing life to physics."

Yet this is an uncomfortably vitalistic view. If molecular biology has taught us anything of lasting significance, it is precisely that life can be reduced to physics, and must be if it is to be properly understood. Surely there is something in human existence—call it mind or consciousness or soul or what have you—that transcends anything physics has shown us to date, but this is a realm in which biology has not yet dared to tread, and one which is not yet touched by the technology of gene splicing. If microorganisms have souls, they have not yet chosen to let us know.

While such metaphysical and moral issues may be debated for some time to come, the legal issue of whether microorganisms can be patented already has been settled, at least for the time being, by no less august a body than the United States Supreme Court, in the so-called Chakrabarty case.

Today, Ananda Chakrabarty is a professor of biochemistry at the University of Illinois at Urbana, but in 1972 he was a research chemist for General Electric, investigating ways in which microorganisms could be induced to degrade toxic chemical wastes. To find genes that were disposed toward this sort of activity, Chakrabarty would collect bacteria from hazardous-waste dumps; some of the

*Morowitz, Harold J. *The Wine of Life*. New York: St. Martin's Press, 1979.

bacteria had developed the ability to devour that waste. In this way he found four different bacteria of the genus *Pseudomonas,* which lived on various components of oil; by taking plasmids from these bacteria (which carried the genes for these functions) and inserting them into a single bacterium, Chakrabarty developed a strain that was capable, in theory at least, of consuming ecologically hazardous oil spills—no small achievement, though the Chakrabarty organism really was too fragile to be considered more than a trial run for other organisms of similar type.

When Chakrabarty attempted to patent this organism in the name of General Electric, he was told by the U. S. Patent Office that living creatures could not be patented, that they were not covered by patent law. GE appealed the case. After eight years it reached the U. S. Supreme Court, which in June of 1980 overturned the Patent Office decision (and thereby affirmed an earlier decision by the Court of Customs and Patent Appeals). Living organisms, according to the Supreme Court, were indeed patentable.

The arguments that led to this decision are of some interest. Patents, according to law, are granted to anyone who "invents or discovers any new and useful process, machine, manufacture, or composition of matter, or any new and useful improvement thereof." Chakrabarty's microorganism clearly was a "new and useful . . . composition of matter"; nothing like it had existed in nature before; but the Patent Office maintained that Congress had expressly excluded living organisms from being patented. In 1930 and in 1970 Congress had passed special extensions of patent law that extended patent protection to new varieties of plants. If Congress had intended that living organisms be patentable, the Patent Office argued, why had they passed special laws to cover plants?

In his majority opinion, Supreme Court Chief Justice Burger countered that in these two cases Congress had intended merely a clarification of existing law, not an extension. "No Committee or Member of Congress," wrote Burger, "expressed the broader view, now urged by the Government, that the terms 'manufacture' or 'composition of matter' exclude living things."

The Supreme Court, however, was careful to draw its opinion on narrow legal grounds, avoiding the murky swamp of ethical and moral considerations that surrounded the gene-splicing issue. Wrote Burger:

To buttress its argument, the Government, with the support of *amicus,* points to grave risks that may be generated by research endeavors such as respondent's. The briefs present a gruesome parade of horribles. Scientists, among them Nobel laureates, are quoted suggesting that genetic research may pose a serious threat to the human race, or, at the very least, that the dangers are far too substantial to permit such research to proceed apace at this time. We are told that genetic research related technological developments may spread pollution and disease, that it may result in a loss of genetic diversity, and that its practice may tend to depreciate the value of human life. These arguments are forcefully, even passionately presented; they remind us that, at times, human ingenuity seems unable to control fully the forces it creates—that with Hamlet, it is sometimes better "to bear those ills we have than fly to others that we know not of."

It is argued that the Court should weigh these potential hazards in considering whether respondent's invention is patentable subject matter under [federal law]. We disagree. . . . We are without competence to entertain these arguments—either to brush them aside as fantasies generated by fear of the unknown, or to act on them. . . .

With this decision, the Patent Office began to process applications for recombinant-DNA-related patents that had been filed prior to the announcement; there were 114 of them in all. Ironically, the first patent issued was not for Chakrabarty's strain of pseudomonas, but for a strain of antibiotic-producing organism developed by researchers at Bristol-Myers. Perhaps even more ironically, General Electric had never had any intention of actually using Chakrabarty's microorganism, even after it was patented; the entire issue had been one of principle, an attempt to test the patent laws, not to obtain a useful patent. The microorganism itself, as noted earlier, was a rather feeble one, more useful for what it proved could be done than for what it actually did. By the time the patent was granted, even the techniques that Chakrabarty had used to create the organism were outdated, superseded some years earlier by recombinant-DNA technology. (Chakrabarty's genetic engineering had involved the use of already existing plasmids rather than gene-spliced hybrids.)

In fact, the Supreme Court decision itself was more useful for the air of legitimacy it conferred on the newborn gene-splicing industry that for any legal advantages it offered. Many industry insiders doubted that the gene splicers would bother to patent their most valuable organisms: Changes in molecular biology were coming too hard and fast. A microorganism patented in 1982, say, might be out of date by 1985, perhaps even before the patent could be processed. And patent law required that an applicant include a full description of the thing being patented, a description that then would be made publicly available. Applying for a patent, then, could involve divulging important trade secrets. Better, perhaps, to keep them secret.

Patents on processes are considered far more important than patents on organisms; if you hold the legal rights on the method you used to create an organism, then you effectively have the rights to that organism (unless, by chance, someone finds a way to make it with a different process). Since much of commercial gene splicing involves marketing only the product of the microorganism rather than the organism itself, the nature of the organism can be effectively kept under wraps.

The patent on the most important process of all—the recombinant-DNA techniques themselves—is held by Stanford University, where most of the original work was done. Oddly, the technique almost was not patented; shortly before the deadline for application, a university administrator pointed out to Stanley Cohen and Herbert Boyer that if they failed to secure the rights to the gene-splicing process they might well damage the future of the biotechnology industry.

This strange bit of logic derives from historical precedent. When Alexander Fleming isolated penicillin mold in the 1920s, he refused to apply for a patent relating to his discovery, believing that such a valuable substance belonged not to any individual or corporation, but to the world. Thus penicillin passed into the public domain—and no corporation wanted to produce it without patent protection. For a decade and a half the public was denied the benefits of penicillin simply because Fleming had failed to secure the rights to it; only World War II and the urgent need for antibiotics offered an incentive for the pharmaceuticals industry to gear up for the fermentation of penicillin, and then only with government help.

Cohen and Boyer were persuaded: It was best that they not allow the recombinant-DNA techniques to pass into the public domain (though it is hard to escape the suspicion that someone at the university recognized that there might be a fair amount of money in the patent). A patent on the techniques was granted late in 1980, and in 1981 Stanford announced the terms under which it would be licensed.

The university requires no licensing fee from those involved in nonprofit research; corporations, on the other hand, are required to pay a $10,000 annual licensing fee and a small percentage of all earnings that accrue from recombinant products.

All things considered, the licensing terms of the Stanford patent are surprisingly lenient; the fees are small, even if they do offer nonexclusive use of the techniques (which is to say that no single company can buy exclusive rights to recombinant-DNA technology).

There has, however, been a certain amount of controversy surrounding the Stanford patent, involving other researchers who feel that their names belong on the patent as well. In response, Stanford's lawyers have asserted that Boyer and Cohen were the "major inspiration and direction" behind the research. It looks as though the Stanford patent will withstand this litigation, though there are those who feel that this sort of squabbling between scientists over potentially lucrative proprietary rights to scientific discoveries may ultimately prove damaging to the American research establishment in general.

They may well be right.

THE CORPORATE FRONTIER

The companies that have sprung up for the commercial exploitation of gene-splicing technology tend to be small, innovative, and closely allied to the university scientists who are at the cutting edge of this segment of molecular biology. (In some instances, these scientists themselves are founding members of the companies, an arrangement that some critics believe may represent an essential conflict of interest between the traditional openness of academic research and the necessary secretiveness of an innovative industry.) Because these companies exist in the shadow of the long-established giants of the pharmaceuticals and chemicals industries, their futures are uncertain, but for the moment they exist at the focus of a great deal of exciting and controversial activity.

The first recombinant product offered for sale commercially was DNA-ligase, one of the enzymes used in gene splicing; it was produced in 1975 by New England BioLabs, a small corporation catering largely to other gene-splicing agencies. Subsequently, a number of other gene-spliced laboratory enzymes have been offered for sale by various companies, but no recombinantly manufactured human proteins have yet come to market. The reason is simple: The FDA requires stringent testing of any chemicals used for medical purposes. And it is the companies that are now preparing to bring such products to market that are receiving the most attention.

Much of this attention has come from Wall Street, where the gene-splicing firms are seen either as the most exciting investments since A T & T or as vastly overrated in terms of earning potential, depending on whom you ask. While the prestigious investment firm of E. F. Hutton has for some time been enthusiastic about the future of commercial gene splicing and (more importantly) gene-splicing stock, other firms believe that the industry will not repay investment within the lifetimes of most investors.

What are these companies? Most of them are located in three places: Massachusetts (near Harvard and Yale), Maryland (near the National Institutes of Health), and California (near Stanford and UCSF). Let's take a brief look at several of the more interesting among them.

Genentech Inc.   Perhaps the most exciting of the small gene-splicing companies, Genentech has managed to stay almost constantly in the public eye through a carefully orchestrated series of announcements concerning its latest research coups: the cloning of human insulin in bacteria, the cloning of interferon in yeast, etc. But perhaps Genentech's greatest coup came when the company offered its stock for public sale on October 14, 1980, and the selling price escalated beyond all expectations. That was the day that Genentech, and the gene-splicing industry in general, became a financial force to be reckoned with.

Genentech was the first gene-splicing firm to "go public"; the anticipation on Wall Street was high from the moment the impending stock sale was announced, but no one seemed to realize just how high. Genentech had intended to price its stock between $25 and $30 a share, but the advance demand was so great that the price eventually was raised to $35 a share, and an extra 100,000 shares were added to the planned 1 million. Shortly after bidding had begun, the price per share reached $89, before settling back to $71.50 by that day's closing.

The performance was astounding. Wall Street was amazed. But many brokers looked askance at this upstart industry. Despite the incredible demand for biotechnology stock at the time of the Genentech offering, the gene-splicing business was regarded by most experienced brokers as a risky venture at best. Though fraught with promise, it had yet to bring forth product; no recombinant merchandise had yet been sold to the public.

There also was a general fear that the enthusiasm over Genentech would create an atmosphere of enthusiasm over genetic engineering stocks in general, driving up the prices of inferior stocks that related in any way to gene splicing. The biggest problem with these stocks, according to one analyst, "is that we are going to have to wait six, seven, or maybe even eight years to see any kind of payoff"—if any payoff was seen at all.

As if to confirm the worst fears of the skeptics, Genentech stock fell to $51 a share by its third day of sale; it then declined steadily for three months until it reached the $36–$45 range, where it has subsequently remained. Not a bad showing, all in all; better, in fact, than initially anticipated. But it was obvious that the initial enthusiasm had been somewhat overblown.

Genentech was founded in 1976 by biologist Herbert Boyer, codiscoverer of the techniques upon which the corporation depends, and investment broker Robert Swanson. In a sense, this rather unlikely partnership underscores the paradox at the center of the entire gene-splicing industry: It is an awkward marriage of pure biology, long considered one of the least lucrative of sciences, and high finance.

The marriage, however, seems to be working; Genentech's impressively successful stock offering made both Boyer and Swanson millionaires—on paper, at least. The corporation has entered into joint ventures with several major corporations, including Hoffman-LaRoche, with whom they intend to market human interferon; Eli Lilly, who will join Genentech in marketing human insulin; and A. B. Kabi, with whom Genentech will market human growth hormone.

To date, Genentech has announced the production of more recombinant-DNA-made products than any other company. These include human insulin, human growth hormone, human interferon, and a number of less glamorous chemicals.

Cetus Corporation   Alone among the front runners in the gene-splicing derby, Cetus began operations some years before recombinant DNA technology became available. Founded by biochemist Ronald Cape and physician Peter J. Farley, Cetus was established in the conviction that there was money in molecular biology, something that no one else at the time seemed to have noticed.

"There had been twenty years of incredible advances in the field," says Cape, "a couple of dozen Nobel prizes awarded, and still not a single practical or commercial application."

While Genentech was founded by a scientist and a businessman, Cetus was founded by two scientists who also are businessmen, individuals who have spent their lives working both in business and in the sciences and who understand both. In September 1980 they decided to follow Genentech's lead and make a public stock offering, amazing investors by offering 5.2 million shares, one of the largest public offerings in Wall Street history. When the stock went on sale in March 1980 it did not cause quite the hysteria that Genentech's offering had generated, but it sold well—closing at $23 a share and raising a total of more than $100 million, an amazing figure considering that Cetus, by its own admission, had lost $3.6 million in the year ending June 30, 1979, and $2.5 million in the next eight months.

By this time, however, financial insiders were less than sanguine about the prospects for gene-splicing stocks in general and Cetus' offering in particular. One investment analyst called the stock "grossly overvalued." The vice president of a West Coast investment house carped that "Cape and Farley are masters at mesmerizing people. I don't say they aren't sincere and dedicated, but unless you are truly long term from the investment standpoint, don't expect to get a payback in your lifetime." In other words, for all the excitement about genetic technology, the payoff is a long way off, and investors expecting to buy a piece of a company with the potential of a new Polaroid or Xerox had best go elsewhere.

Cape and Farley disagree. "Even the critics admit that we will have a profound impact in a number of different industries," Cape says.

Though their accomplishments have been more low-key than those of Genentech, Cetus' list of accomplishments is impressive: They have located a bacterium that manufactures fructose, a form of sugar one and a half times as sweet as ordinary table sugar and that already has caught the fancy of the soft-drink industry; and they have cloned interferon in *Bacillus subtilus*, a hardier and more efficient microorganism than *E. coli*. Currently they are developing microorganisms to produce chemicals that may prove valuable in the plastics industry.

According to Peter Farley, Cetus has no intention of concentrating strictly on pharmaceuticals, as Genentech has done. "The philosophy here is to take biology wherever it will lead us."

In addition to its publicly owned stock, Cetus is partially owned by Standard Oil of California, National Distillers, and Standard Oil of Indiana.

Biogen   Based in Switzerland, Biogen first came to public attention on January 16, 1980, when representatives of the corporation announced that they recently had produced biologically active human leukocyte interferon in bacteria. Though this was not the first time that interferon had been produced in bacteria (the Japanese had already cloned fibroblast interferon the year before), the event was given major coverage by the media. Stock in Schering-Plough, part owner of Biogen, rose eight points. Biogen became well known virtually overnight.

Biogen's announcement was less than the breakthrough it may have seemed at first, but still it was an important announcement. Interferon clearly was the glamor product for recombinant-DNA production, and anyone who had an edge, however small, at being the first to market it was in a strong position among biotechnology firms.

In order to isolate a copy of the interferon gene, Biogen's resident molecular biologist, Charles Weissmann, had performed a meticulous search, lasting many months, for a piece of mRNA coding for interferon. Knowing that if he took sufficient cytoplasm from white blood cells, he would eventually come across at least one copy of the gene, Weissmann created 15,000 recombinant bacteria and patiently sorted through them until he found the right one. It was a triumph of scientific forbearance.

In addition to Schering-Plough, which owns 16 percent of Biogen, International Nickel owns 24 percent of the corporation.

These are only a few of the companies to emerge over the past several years; a complete list would take another book. Within a few years, this picture of the biotechnology industry may be completely out of date, so rapid is change within the field. Most industry watchers see a great deal of money in gene splicing's future—but it will come from only a few companies; the rest probably will be lost in the wash. One current problem is duplication of effort: Too many of these firms are tackling the same problems. According to one analyst, even Genentech, the darling of the industry, has yet to produce a single biological product that is not being researched by at least one other company. "There is no way of knowing who is going to have the most successful product."

Perhaps the greatest threat to these tiny corporations are the industry giants, who have until now waited on the sidelines. Companies like Dupont, Monsanto, and Dow have the knowledge and the staff support to do everything that the small companies are doing right now and to do it better than any of these companies can. And some of these big corporations are showing signs of becoming involved.

If they do become involved, few if any of the small companies will survive—or, at least, this is the common wisdom. Most likely, they simply will be absorbed by the giants; it would be a lot cheaper for Dow or Dupont, say, simply to buy out Genentech or Cetus instead of spending the money to create their own biotechnology R&D departments. Why make your own when you can buy one off the rack?

AUTOMATED EVOLUTION

As biology and industry merge, the life sciences will come more and more to take on the trappings of industry. One of these trappings, in some ways long overdue, is automation.

In 1981, two companies (Vega Biochemicals of Tucson, Arizona, and Bio Logicals of Toronto, Canada) offered for sale automated gene synthesis devices (popularly known as gene machines), and as many as fifty more companies may be waiting in the wings with competing models. Simply put, these devices make genes, and they make them to the biologist's order. All the user need do is add nucleotides and chemical reagents at one end, punch the proper order of nucleotides for the desired gene into the attached microcomputer, and (after perhaps twelve hours of synthesis) out pops the gene, which then can be spliced into a recombinant bacterium.

Sound amazing? Actually, gene synthesis has become such a well-established process that it represents little more than time-consuming busywork to a molecular biologist, on the order of washing dirty test tubes or putting labels on Petri dishes. Yet the glorified bottle washers who do the work of gene synthesis are often molecular biologists with Ph.D.'s who have much more important work to do. "The situation," according to biochemist Thomas G. Mysiewicz, "is somewhat as if integrated circuit manufacturers put their degreed electrical engineers and circuit designers to work on the production line. It's hardly the way to run a business."

Now all the biologist need do is leave the gene machine running when he or she leaves work at night and the completed gene will be ready the following morning, freeing the biologist's time for more demanding work.

How do these gene machines operate? Inside a short metal column is affixed a so-called solid support matrix, to which nucleotides can be physically attached (much as enzymes are attached to glass and plastic beads in some fermentation processes). Chemical solutions, containing either nucleotides or reagents, then can be poured into this column, to react in prearranged ways.

The nucleotide attached to the support matrix is the first nucleotide in the desired gene sequence. Above the column where the

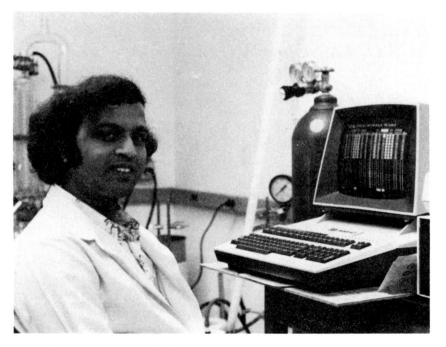

Here Krishna Jayarama, Ph.D., of Genex Corporation waits on his gene machine to put together the genes he wants. *(courtesy Genex Corporation)*

nucleotide is fixed sit eight metal reservoirs (actually, small cylindrical containers) containing supplies of DNA's four different nucleotides (adenine, guanine, cystosine, and thymine) and four reagents, each in separate containers. When the second nucleotide in the desired sequence is needed, it is released into the support column, where it can attach itself to the first nucleotide.

This is a trickier business than it might at first seem. When two nucleotides in a DNA strand attach to one another, they must attach themselves in the proper orientation, with the so-called 3' (pronounced "three prime") end of one nucleotide attaching itself to the 5' ("five prime") end of the preceding nucleotide. Otherwise, one nucleotide might wind up backward in relation to the other.

The first nucleotide in the sequence is attached to the support matrix by its 3' end, so that only the 5' end is available for bonding. But the second nucleotide, floating free within the column, can attach itself in any orientation; therefore this nucleotide is specially treated before insertion so that its 5' end is capped off by a small molecule and is not free to react. Attachment has to occur at the 3' end.

Once attachment has occurred, a chemical reagent is introduced into the column to remove the molecular cap from the 5' end. Then the next nucleotide is introduced into the column, also with its 5' end capped, and so forth. Eventually a chain of nucleotides builds up within the column, the sequence having been entirely specified in advance.

One problem with this system is that, to date, only chains up to twenty nucleotides in length (called "oligonucleotides") can be synthesized at a time; these short chains then can be linked together by nonautomated methods to form entire genes.

What makes this process particularly interesting is that it represents a kind of wedding of computers and genetics, a marriage that eventually may bear some interesting offspring. It is now becoming feasible, for instance, to design on computers the genes for organic molecules that never have existed previously in nature.

Because biologists now have some understanding of the ways in which enzymes interact with other molecules, and the ways in which enzymes themselves form out of strings of amino acids, it is possible to model these processes graphically on a computer, with the results displayed on a video monitor. The biologist then can specify the amino-acid composition of a given enzyme and the atomic composition of the molecules with which he wishes that enzyme to interact.

The computer then constructs an image of those molecules and presents an animated representation of their interactions.

In this way, any enzyme that can be imagined can be designed on a computer, given sufficient information about the molecules with which that enzyme will interact. The computer then can specify the order of nucleotides in the gene that would code for that enzyme; the gene then can be synthesized and inserted into a recombinant bacterium. Perhaps the same computer used to design the molecules could monitor the gene-synthesizing process. This might be the ultimate form of computer printout: The specifications for an imaginary molecule are put in one end and the molecule itself comes out the other.

Thus will computers come to the aid of biologists in the not very distant future. According to scientists at the small Rockville, Maryland, firm of EMV Associates, biologists soon may be able to return the favor.

It is possible, says EMV's president, James McAlear, that gene-splicing technology eventually might create microorganisms capable of producing highly miniaturized microprocessors constructed out of organic molecules. If a gene could be designed that specified for molecular electronic circuits, thousands of infinitesimally tiny computers could be fermented out of vats of bacteria.

Kevin Ulmer, of Genex Corporation in Bethesda, Maryland, agrees: "The ultimate scenario is to develop a complete genetic code for the computer that would function as a virus does, but instead of producing more virus, it would assemble a fully operational computer inside a cell."

Such computers would not only be extremely small (and therefore capable of extremely rapid calculations, since the speed at which a computer works is largely determined by the distance that signals must travel through its circuits), but also it might be possible to interface it directly to the human brain, bypassing ordinary forms of input and output. Instead of placing information into the computer via a keyboard or punch card, such a biological computer might accept data directly from the nervous system. In turn, it could output information straight to the brain. Such biological computers might even serve as replacements for damaged nerve tissue, offering eyesight for the blind or hearing for the deaf.

How close are we to this sort of technology? Probably not very close. But in the rapidly advancing field of molecular biology it is hard to say when the breakthroughs necessary for such technology may come about. In ten or twenty years this sort of technology might seem not just feasible but even ordinary. And who knows what might lie on the horizon by then?

Perhaps the biological engineering of the human organism itself.

PART FOUR

The Engineers of Life

NEW, IMPROVED HUMANS?

At the height of the recombinant-DNA controversy, most of the debate turned on the question of risk. Were recombinant organisms safe? Would they represent a danger to humans or to the environment? Were molecular biologists prepared to take proper precautions in this work?

By no one's assessment were these easy questions, but those engaged in the debate at least could argue them with the assurance that some kind of answers eventually would be found, for better or worse. To some degree these questions were indeed answered, though hardly to everyone's satisfaction.

There were, however, other questions raised during the debate that were not so easily addressed, because they turned on less tangible issues—moral questions, questions of propriety and responsibility. If science had made feasible the direct alteration of the subtlest aspects of life, how much right did we have to follow where science led?

For some, the darkest implication of gene splicing was that someday soon it might be possible to tamper not only with the genes of bacteria but also with the genes of men and women; for others, this was gene splicing's greatest promise. With the death of vitalism, and the advent of an increasingly mechanistic view of life, biology— and, by extension, medical science—was coming to take on some of the attributes of a television repair shop. If only we could understand the mechanism of life as well as we understood the mechanism of a television set, it might be possible to take the human body into the shop, disassemble it, and put it back together in greatly improved fashion.

But who is to define what represents an improvement? Should

117

impersonal science be allowed to tamper with this most personal of possessions—the genetic heritage of the individual?

These were not the sorts of questions that could be answered at press conferences, or even in academic debates. Most scientists shrugged them aside by muttering vaguely as to how we certainly were nowhere near able to engineer the human genome, which is true, and how we wouldn't be able to do it within the lifetime of anybody listening to the question, which is less obviously so.

The fact is, while we are far from being able to engineer the human chromosome (or the chromosome of any organism more advanced than a cauliflower), we have nonetheless amassed a considerable body of knowledge as to how such engineering might be done, pending a number of breakthroughs necessary to make it possible. And, given the breathtaking rapidity with which molecular biology has advanced under the impetus of recombinant-DNA research, who is to say that those breakthroughs are not just around the corner? Hadn't we best be prepared for them? If the question of genetic engineering is to be debated, perhaps the time is now, while our heads still are reasonably cool, rather than the day after tomorrow, when we may wake to find that the technology already is in place and that the first act of human genetic engineering has already been performed.

In a lecture given in 1977 at the Brookhaven Symposium on Biology, Charles A. Thomas, a biochemist at Harvard Medical School, attempted to lay to rest the specter of human genetic engineering that had haunted molecular biology since Asilomar. To this end, Thomas presented a list of problems that would have to be overcome before even a simple, effective alteration of a single defective human gene—a process sometimes referred to as gene therapy—could be successfully carried out. His conclusions? "I would not bet one penny that gene therapy will ever be a practical reality, and all the talk to the contrary is wrong and misleads the public. Talk about genetic engineering and the manipulation of human genes is thus irresponsible."

Thomas' arguments revolved mainly around the difficulties of inserting a new gene where it could be expected to do some good, getting the gene to express itself once it has been inserted, and keeping that gene functioning properly for the rest of the patient's life. Clearly, Thomas felt that these difficulties were effectively insurmountable.

But are they? Even in the half decade since Thomas' speech his reasoning has come to seem less firmly grounded: As we shall see in a moment, genes have been effectively transferred between rabbits and mice; natural gene-transfer mechanisms have been discovered within the cells of higher organisms; and at least one scientific team has attempted human gene therapy using recombinant-DNA techniques, in a highly controversial series of experiments.

Why would we want to perform this sort of gene therapy? Well, in its simplest form, such technology might be used to cure inherited diseases, of the sort in which the victim has received a defective gene for a single enzyme. If that gene could be replaced, even in a few cells of the victim's body, some measure of relief—perhaps even a complete cure—might be granted.

Sickle-cell anemia is one such disease. In this condition the sufferer has received from both parents a defective gene for the hemoglobin molecule, which transports oxygen through the blood. The defective gene causes the blood cells to take on a characteristic "sickle" shape; the body's oxygen-transport system breaks down, leading to debilitating illness and often death.

Beta thalassemia is a condition closely related to sickle-cell anemia, except that beta thalassemia renders the body scarcely able to make hemoglobin at all. Blood cells are manufactured in the bone marrow, and in the thalassemia victim the bones swell to grotesque size in an attempt to produce enough hemoglobin to keep the body alive. Blood transfusions can be used to prolong the victim's life, but few sufferers live beyond their twenties.

Other diseases would be amenable to such treatment as well; one estimate puts the number of such single-gene diseases at 2,500. Gene therapy might offer a cure for diabetes, for instance, or hemophilia.

But how to get the gene into the victim's body? Ah, there's the rub.

There is already a crude method of genetic engineering available, though it offers only limited solace to those sufferers now alive: Carriers of the defective gene can refuse to have children by other carriers. Most genetic diseases are recessive traits—that is, they express themselves only when an individual receives genes for the condition from both parents. Those who carry only a single gene rarely exhibit symptoms of the disease, but they can pass on these genes to future generations. Such diseases are comparatively rare, because the odds are slim that two carriers of the defective gene will mate (which is one of the reasons the diseases were not bred out of the population thousands of years ago by natural selection; evolution cannot act on a recessive gene until it expresses itself).

If a carrier can go through life without ever realizing that he or she has the gene for, say, sickle-cell anemia, how can the carrier know that he or she is a carrier—and thereby avoid having children by another carrier?

Genetic screening is one alternative; it is offered at a number of clinics around the country. Before having a child, a couple can ascertain whether or not they carry any defective genes and whether there is any change of passing them on; then they can plan their family accordingly. If they decide to take the risk and conceive a child despite a perceived genetic hazard, the pregnant mother may choose to undergo amniocentesis, a process by which the fluid from her womb can be analyzed for clues as to the fetus' genetic heritage. If the child has inherited the condition, then the couple can decide whether they wish to bring the pregnancy to term.

This raises the question of what to do in the event the unborn child is suffering from some debilitating condition. Many couples opt for abortion. Obviously, this is an inadequate solution—and a controversial one—but there is, at present, no real alternative if the defective gene is not to be perpetuated. Perhaps less obviously, genetic screening offers the possibility that an individual found to be carrying the gene for a disease might be stigmatized by that knowledge. In some communities, those found to be carrying genes for sickle-cell anemia or beta thalassemia have been ostracized by their neighbors as somehow inferior; much of the problem has grown out of the tenuous grasp that most people have of genetic concepts, but this still is a real concern.

Whatever the benefits and drawbacks of genetic screening, it still is only a stopgap form of gene therapy. It in no way cures the real problem: the genes themselves.

HOW TO REMAKE MAN (OR WOMAN)

How would we go about actually entering the human chromosome and making changes?

First we have to have some idea of what the chromosome is like, what genes are on it, what proteins they code for, and what the nucleotide sequences are. These are not easy things to find out; nucleotides are too small for direct examination, so they must be analyzed indirectly.

The early genetecists solved this problem by applying mathematical analysis to the ways in which genetic traits were inherited by organisms such as drosophila and *E. coli*. Early genetic researchers understood that genes were strung together on chromosomes and therefore that certain genes usually were inherited in conjunction with certain other genes, because they reside on the same chromosome.

Usually, but not always. Though many genes tended to be inherited together, few genes were invariably inherited together. This resulted from a phenomenon called "crossing over": When chromosomes in the sex cells of an organism—the cells that are passed on to further generations—divide, sometimes they become intertwined. Large segments of DNA can be exchanged between chromosomes, the genes on these segments actually passing from molecule to molecule. Thus two genes on the same chromosome can be separated when crossover occurs, if one gene crosses over and the other does not.

This was of value to geneticists because it offered a way of mapping the chromosome, determining the order in which genes were arrayed along its length. Two genes that lay near to one another on the chromosome would almost always be inherited together, because the likelihood that a crossover would occur between them was small. On the other hand, two genes at a great distance from one another on the chromosome—at opposite ends, say—often would be inherited separately.

In this way geneticists drew up complex gene maps showing where on the chromosomes of various organisms the genes for certain characteristics fell. This told us little, however, about the specific enzymes coded for by those genes.

Genetic techniques have advanced considerably in the past half century. Now, if we want to know what protein is coded for by a specific stretch of DNA, we simply splice that DNA into a plasmid and insert it into a bacterium, where we can examine the molecules it produces, assuming we can persuade it to express itself. Suppose, however, that we want to know the order of nucleotides on that gene?

We might use a method called *electrophoresis*, in which we

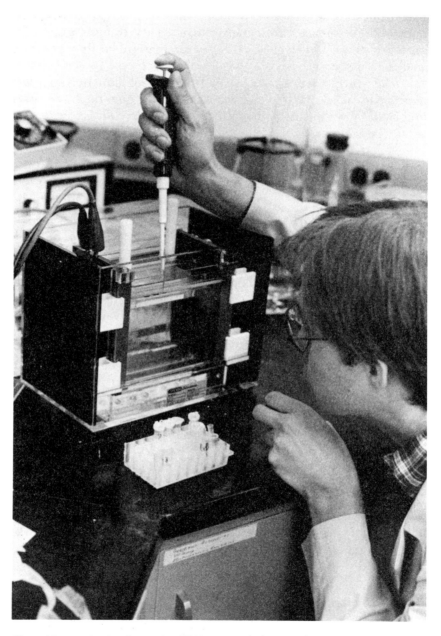

One of the methods of mapping DNA is electrophoresis. Dave Anderson is preparing to separate fragments of DNA electrophoretically. Note the electrodes on the left side of the chamber. *(courtesy Genex Corporation)*

would fragment a molecule of DNA into a number of pieces by exposing it to a chemical reagent—specifically, a reagent that cuts the molecule at the occurrence of a specific base and only at that base. Then we can place these fragments on a gel-covered sheet and expose them to an electric field. The field will cause the molecules to shift position, but the heavier molecules will shift farther than the lighter molecules. As a result, the DNA fragments will be neatly distributed across the gel by weight—the long, heavy molecules at one end, and the short, light molecules at the other end.

This tells us the precise size of each fragment. Because the reagent to which the DNA was exposed cut the molecule only at a specific base, we therefore know which base was at the position where the molecule was cut. If we have fragments three, eight, fourteen, and twenty molecules long, we know that the same base—adenine, say—was at each of these positions in the unbroken molecule. Then, if we expose other, identical molecules to reagents that cut the chain at other bases, we will know where on the chain those bases occur, and gradually we can build up a picture of the base sequence on the molecule as a whole.

There are other ways of doing this, of course. One involves making copies of the DNA using an enzyme that transcribes only three of the four nucleotides, thereby halting transcription whenever the fourth nucleotide occurs. The fragmented copies then can be sorted in the fashion described above.

Using such methods, we can sequence as many as 200 to 300 bases in a single experiment, though sequencing an entire plasmid containing thousands of bases still might take months of exacting work.

Fortunately, these processes now have been automated. Gene-sequencing machines have been developed that can do this work rapidly, with minimal human intervention, removing a nucleotide at a time from the molecule and analyzing it chemically.

Yet the human genome is very large—from 20,000 to 100,000 separate genes. At the Sixth International Workshop on Human Gene Mapping, held in Oslo in 1981, it was estimated that perhaps several hundred genes on the human chromosomes had been precisely mapped as to location and perhaps a hundred of them had been sequenced in their entirety, a significant number but only the tiniest fraction of the whole.

Though it may not tell us much about the specific sequence of nucleotides on the genes, plans now are under way for an ambitious,

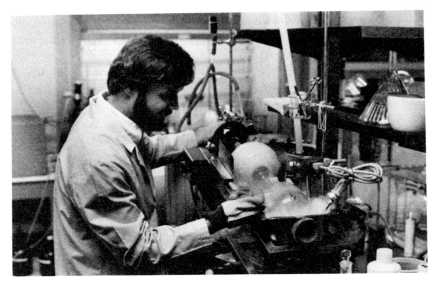

A DNA probe is another way to find out the identity of an anonymous fragment of DNA. This is a mix-and-match process. Expose your unknown to a known fragment of DNA. If it attaches, you know exactly what you have; if it doesn't, you've ruled out one possibility. Here a technician evaporates a DNA probe as part of a nucleic-acid synthesis. *(courtesy Genex Corporation)*

government-sponsored project in which all of the proteins in the human body would be mapped on thousands of sheets of photographic paper. Through electrophoretic techniques, the protein content of each different kind of cell in the body could be mapped. The resulting file of protein data could be used as a standard reference for genetic diseases. A patient suspected of carrying faulty genes could have charts of his or her own cells compared to standard charts, to detect deviations in protein content.

Such a project, involving the cataloging of something like 50,000 protein molecules, would be a major undertaking. Biologist Norman Anderson estimates that the project could be carried out in five years for $350 million. Though several congressmen are interested in the project, such a high level of funding is unlikely, though it is possible that $10 million a year eventually may be allotted for the project.

It will be some time before we know the total nucleotide composition of the human genome. But even before we have a complete understanding of human genetics, it may be possible to correct specific genetic defects in human cells through the use of recombinant DNA techniques—and to reinsert these corrected cells back into the human body.

In fact, it already has been done.

THE FIRST TIME

We know that it works with mice. In 1981 a team of researchers at Ohio University announced that they had inserted a rabbit gene into the body of a mouse. Not only had they gotten the gene into the mouse, but also it had functioned there—and the mouse had passed it on to its offspring.

The gene, however, was introduced into the mouse shortly after conception; in fact, it had been injected into the freshly fertilized mouse egg, where it then multiplied as the egg multiplied, until there were copies of the gene in every cell of the adult mouse.

The team of researchers, headed by biologist Thomas Wagner, spliced a rabbit hemoglobin gene (along with a stretch of DNA that they believed carried the control sequence for the gene) into a bacterial plasmid, then injected copies of the plasmid into 312 mouse eggs. A total of 211 of the fertilized eggs were transplanted into foster mothers; 46 of the eggs eventually developed into live mice.

"We think 15 to 20 percent of the mice were producing some level of the protein [coded for by the gene]," said Wagner later. When two of the mice were mated, they produced eight offspring, at least five of which also were producing the protein.

It was not the first time foreign genes had been inserted into mouse embryos, but it certainly had produced the most spectacular results. Could similar experiments be attempted on human beings?

There's no technical reason why they couldn't; *in vitro* fertilization—popularly known as the "test-tube baby" process—offers a chance for selected genes to be injected into a human embryo before that embryo is implanted into the mother's womb. But it will be a long time before we have a solid enough grasp of such genetic-engineering processes to risk trying them on an unborn child. Besides, gene therapy is a process required more urgently by those who have already been born.

Dr. Martin J. Cline, former chief of the division of hematology and oncology at UCLA, also had attempted to insert new genes into

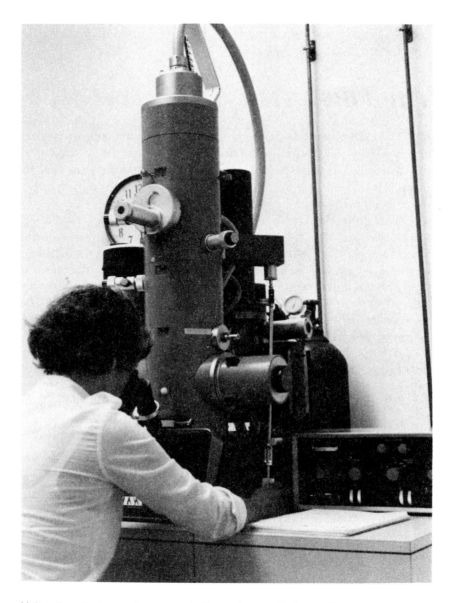

Using the electron microscope is the only way that genetic
engineers get a peek at what they're working with . . .

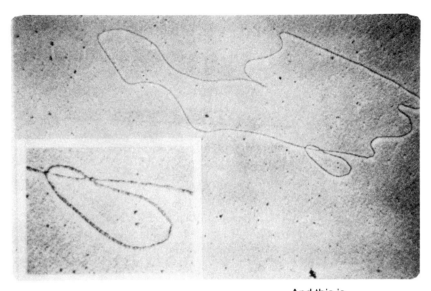

... And this is
what they're working with—DNA. Here you see a molecule of
heteroduplex DNA that's had a gene attached (see insert)
magnified 80,000 times. *(courtesy Genex Corporation)*

mice, though Cline's interest lay in replacing the bone-marrow cells
of living mice; his technique involved the preparation of genetically
altered bone-marrow cells which had been given genes conferring
resistance to a drug called methotrexate, which normally kills
bone-marrow cells. Upon injecting the cells into the bone marrow of
living mice, he then killed the rest of the bone marrow with
methotrexate. In the absence of competition from other cells, the
new methotrexate-resistant cells thrived and eventually replaced
the old bone marrow.

Cline envisioned applications of this technique for victims of
diseases such as sickle-cell anemia and beta thalassemia, both of
which result from defects in the hemoglobin genes of the bone
marrow. It was Cline's conviction that it was time to begin experi-
menting, however tentatively, on human beings.

In 1979 Cline, a specialist in blood disorders, began applying
for permission to perform gene-therapy experiments on human
patients—in Los Angeles, in Israel, and in Italy. The Los Angeles
application became bogged down in red tape, so Cline and a team of
doctors went overseas, where they worked with two young
women—one in Israel, the other in Italy—suffering from beta
thalassemia. Cline took bone-marrow cells from both young women
and incubated the cells together with hemoglobin genes purified out
of recombinantly engineered bacteria. Some of the bone-marrow

cells took up the new genes; Cline then injected them back into the bone marrow of the patients.

Theoretically, at least some of the engineered marrow cells could be expected to take up residence in the women's bone marrow; none of the cells, however, began producing functional hemoglobin, at least not enough to improve the condition of the patients. Cline returned to the United States.

When he did, he ran into a heavy storm of criticism. Apparently he had neglected to get anyone's permission to perform the experiments using recombinant-DNA techniques—and this was at a time when the NIH guidelines still were in full effect. Furthermore, Cline was accused of having broken federal regulations on human experimentation. Within days, Cline had resigned his position as division chief. In 1981 he was disciplined by the NIH, which found him in violation of its regulations.

Cline was repentant. "I greatly regret my decision to proceed with the use of recombinant molecules without first obtaining permission from the appropriate committees."

Cline's work was premature, yes, but there is little doubt that others eventually will follow in his footsteps.

From the viewpoint of those investigators the most important aspect of Cline's experiments will be that they did not work. Why didn't they?

The main problem is that too little is known about human gene expression; genes transplanted into human cells through roundabout gene-therapy techniques will not be easily persuaded to produce their recombinant products.

Cline's bone-marrow transplant method also is rather limited as a method of gene therapy. It may prove an effective treatment for diseases of the blood, but other conditions may be beyond its reach. Is there any way, short of transplanting genes into fertilized egg cells, that we can perform gene therapy on cells other than those in the bone marrow?

Perhaps. And the key to this ultimate gene therapy may well turn out to be that ancient messenger of disease, the virus.

JUMPING GENES

Viruses, as we have seen in earlier sections, are nature's way of getting foreign DNA into cells. They function much like miniature hypodermic needles, injecting their DNA into the cell, where it sometimes becomes part of the chromosome.

Looked at one way, this is a form of natural genetic engineering; the very presence of viral DNA changes the genetic makeup of the cell. And in many instances the viral DNA is capable of expressing itself within the cell.

This fact has not escaped the attention of potential gene therapists. Even before recombinant DNA came onto the scene, researchers were paying a great deal of attention to an organism called the Shope papilloma virus, which was known to carry the gene for arginase, an enzyme that breaks down the amino acid arginine. A rare disease exists in which the victim's cells are unable to synthesize their own arginase; in an experiment, two victims of this disease were exposed to the Shope papilloma virus, in the hope that it would contribute the arginase gene to their cells and reverse the condition.

The experiment was partially successful. The deficient cells began to produce arginase, but the change either was too little or too late to stop the course of the disease. Nonetheless, the arginase gene had found its way from the virus into the cells; a crude sort of genetic engineering had taken place.

Can viruses, then, serve as vectors for gene therapy? The potential exists, yes, though it has been little explored; certainly no other obvious method of reaching large numbers of cells in an adult human being has presented itself. Furthermore, there is at least some indication that genes within the cells of higher organisms already may travel from cell to cell (and perhaps even from body to body) on viruses.

As scientists have become increasingly aware of the surprisingly large amounts of useless DNA in the eukaryotic genome (the "selfish DNA" we discussed earlier), they also have seen signs that certain genes have the ability to jump from place to place on chromosomes and even from chromosome to chromosome within the nucleus of the cell. This phenomenon was first noticed in the genes of corn more than thirty years ago by Dr. Barbara McClintock at Cold Spring Harbor Laboratory, but only recently has it become obvious that it is not restricted to one or two different kinds of organism; it appears to be a common process.

These "jumping genes," called *transposons*, apparently can synthesize enzymes that literally slice them free of the chromosome, so they can reattach themselves elsewhere in the cell; they even may be able to incorporate themselves into the genes of invading viruses.

The debate continues even now. *(courtesy National Institutes of Health)*

Such a hitchhiking transposon then would be able to escape the cell and travel almost at will through the intercellular fluid, perhaps eventually becoming part of another cell, even another organism. It even has been suggested that this is how cancer genes may travel, in certain species at least; a cancer gene in one cell may be "accidentally" picked up by a virus and carried to other cells, sometimes to other individuals, on occasion to members of other species.

It is possible, then, that human scientists are Johnnies-come-lately to the techniques of genetic engineering; perhaps the genes themselves have been using these techniques for millions of years to modify and rearrange themselves, to spread the genetic wealth—and sometimes the genetic poverty—across a wide spectrum of recipients. This may even be one of the mysterious ways in which evolution functions, a method of promoting the great and sudden alterations of species that have so baffled paleontologists and anthropologists, a form of drastic chromosomal change that could explain how apes could evolve in only a few short millions of years into a species capable of taking over its own evolutionary guidance.

For we do in fact have the tools for taking evolution into our own hands and guiding it down new channels, or at least we are on the verge of having such tools. No longer need we be at the mercy of blind processes of change; like farmers performing sophisticated genetic selections on their crops, we can take the sightless process of natural selection and give it eyes.

But first we must clear our own vision. The fear of those who saw in recombinant-DNA techniques the first inroads of genuine human genetic engineering was that this potential would be used irresponsibly, mindlessly, that perhaps humanity is not ready to engineer its own genes. They may be right. But once we have begun on such a road it is difficult to turn back, and perhaps it would be foolish to do so. We stand on the edge of magnificent things, but like all great scientific advances genetic engineering can be abused—and probably will be. It is time now to decide what we wish to do with this ability—with gene therapy, with genetic engineering. It is time to debate our options before we find that they are options no longer. Gene therapy, whether it comes tomorrow or a century from now, offers us the chance to cure diseases that no one ever thought could be cured. It offers us the ability to take control of our own genetic destiny. It even may make us immortal if we want to be. Now is the time to assess the prices we will have to pay.

Now that we can rewrite the genetic code, what will we say? What indeed?

Index

A

active sites, 9
adenine, 13
Agrobacter, 98
amino acids, 9
anaerobic microorganisms, 77
antibodies, 64
antibiotics, 41
Aristotle, 4
Asilomar Conference, 47, (II) 54
atoms, 7
autoimmune diseases, 66
Avery, Oswald, 12

B

bacterial plasmid, 42
bacteriophage, 34
Baltimore, David, 50, 67, 70
base, 13
base pairing, 13
beta thalassemia, 119
Berg, Paul, 32, 35, 67, 68
biogen, 110
bonds, 7
Boyer, Herbert, 38, 48, 108
Brenner, Sydney, 55
Bulletin of the Atomic Scientist,
The, 62

C

Cline, Martin J., 125
Cape, Ronald, 108

catalysis, 9
cell, 1
cell fusion, 88
cellulase, 64
Cetus Corporation, 108
Chakrabarty, Ananda, 102
chemical reactions, 7
chromosome, 3, 6
codon, 21
Cohen, Stanley, 37, 68
commercial genetics, 101
complementary base pairing, 42
containment lab, 63
Crick, Francis, 12
"crossing-over," 121
cytosine, 13

D

daughter molecules, 19
Davis, Bernard D., 62
Dawkins, Richard, 72
deoxyribonucleic acid (DNA), 11
DNA probe, *124*
Dubos, René, 29
Dworkin, Roger, 57

E

electrophoresis, 121, *122*
EMV Associates, 114
enzymes, 9

Escherichia coli bacterium, 30, 40, *41*, 46, 56–57, 59, 62, 64, 67, 109
eukaryotic cells, 36
evaporator, *iv*

F

Farley, Peter J., 108
fermentation, 79, 94
Fleming, Alexander, 81, 105

G

gene synthesizers (gene machines), 111
genes, 3, 5
genetic code, 13, 23–24
genetic diseases, 119
genetic engineering
  in agriculture, 96
  in food processing, 96
  in fuel production, 95
  in mining, 95
  in pollution control, 95
genetic screening, 120
genetics, 5
Genentech, 107
Genex Corporation, 114
Goodfield, June, 52
Gordon Research Conference on Nucleic Acids, 48
Griffith, Russell, 12
guanine, 13

H

Handler, Philip, 50
Hayflick's limit, 88
HLA antigens, 89
hormones in the brain, 90
human growth hormone, 85
hybridoma, 89

I

insulin, 82
interferon, 83
International Conference on Recombinant DNA Molecules, 54
introns, 71
Isaacs, Alick, 83

J

Jacob, François, 25
Jayarama, Krishna, 112

K

Kennedy, Senator Edward, 68
Kohler, Georges, 89
Kornberg, Arthur, 29

L

Lewis, Herman, 50
life, 2
Lindemann, Jean, 83

M

McAlear, James, 114
McClintock, Barbara, 129
Mendel, Gregor, 4
Mertz, Janet, 45
messenger RNA, 19
Miescher, Friedrich, 11
Milstein, Cesar, 89
molecules, 7, (model) 13
molecular spine, 13
Monad, Jacques, 25
monoclonal antibodies, 87
monomers, 8
Morgan, Thomas Hunt, 5
Morowitz, Harold J., 102
mutations, 17–18
Mysiewicz, Thomas G., 112